Steps in Physics

John Byron

Deputy Headmaster, Hutton School, Lancashire
formerly Head of Science, Rainford High School, St. Helens

Oxford University Press 1979

Oxford University Press
Walton Street,
Oxford OX2 6DP

© Oxford University Press 1979

Oxford London Glasgow New York Toronto
Melbourne Wellington Ibadan Nairobi Dar es
Salaam Lusaka Cape Town Kuala Lumpur
Singapore Jakarta Hong Kong Tokyo Delhi
Bombay Calcutta Madras Karachi

© Oxford University Press, 1979

ISBN 0 19 914058 8

Printed in Great Britain
by William Clowes & Sons Limited
Beccles and London

Introduction

What Science is about. Science is about asking questions. You can ask questions about absolutely anything – you ask scientific questions when you can do *experiments* to find the answer to them.

The value of your answer will depend upon the quality of your question. Scientists have to be very careful to ask questions from which the answers can be trusted.

What physics is about. Physics asks questions about *things, forces* and *energy.* What force makes a thing like this book move if it falls? Where does the energy come from to make this happen?

What Steps in Physics is about. This book provides answers to some of the huge numbers of questions that physicists have asked. The information in this book gives you a background for the experiments you will do. Your experiments should answer questions that you have thought up for yourself, or your teacher has thought up for you. Enjoy asking questions – use this book if you want to find answers quickly!

How to use this book. This book is organized so that you can find out things quickly and easily. For this reason, it is written in two-page units. Each unit is on a topic you are likely to study. To find out what you want:

Use the contents list. If you want to find out about a large topic, look through the contents list – you should find it there. The units are grouped into chapters, so you can find your way around easily.

Use the index. If you just want to find out about one small thing, then look up that word in the index at the back of the book. After each word is the page number where that word appears.

Use the questions. You may use the book for homework, or for revision before an exam. At the end of each topic, there are "exercises". These will help you make sure that you really have understood the unit. At the end of the chapters, there are actual exam questions. These should give you an idea of the sort of questions that are asked in exams. To help you, the answers are at the back of the book.

Contents

Mass, force and weight

Mass, force, and weight and words that have special meanings in Physics. It is important to understand exactly what they mean – they will be used many times throughout this book.

Mass

The mass of an object is simply a measure of how much material there is in it. Large objects (such as the Earth) have a large mass, and small objects (such as a feather) a small mass. The amount of mass in an object stays the same, no matter where it is. Mass is measured in *kilograms* (kg).

The three photographs show an astronaut in different places. On Earth, near an orbiting space craft, and on the Moon, he has a mass of 75 kg.

The mass of the astronaut is the same no matter where he is. To summarize:

Mass is a measure of the amount of matter in an object. Mass is measured in kilograms (kg).

The mass of the astronaut in space is 75 kg.

The mass of the astronaut on Earth is 75 kg.

The mass of the astronaut on the moon is 75 kg.

Force

The force on an object is the push or pull on it. There are many types of forces: for example there are elastic forces, magnetic forces, and electric forces. The strength of a force is measured in newtons (N). An idea of the size of the newton is given by the following examples: A force of about 4 N is needed to lift this book. A force of about 1000 N is needed to push a car which will not start. The engines of the space shuttle will produce a force of about 30 000 000 N.

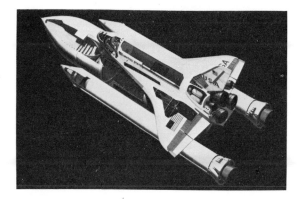

Weight

Mass has one property which is still not properly understood. For some reason there is a force of attraction between any two masses which causes them to pull towards one another. This force is called the force of gravity. The force of gravity depends on the size of the two masses and their distance apart. The force is large when the masses are big and close together.

One example of the force of gravity between masses is the way in which the Earth pulls all objects towards it. The force pulling this book towards the Earth is about 4 N. The force of gravity on this book is 4 N. The force of gravity on this book is called its *weight*. The weight of this book is 4 N. Since weights are forces, they too are measured in newtons.

Unlike mass, the weight of an object changes depending on where it is! A mass of 1 kg on Earth has a weight of about 10 N. An astronaut of mass 75 kg has a weight on Earth of about 750 N. After he takes off, even though his mass stays the same at 75 kg, his weight changes. In orbit he floats – he is

"weightless". The motion of the spacecraft going around the Earth balances the force of gravity. When the astronaut lands on the Moon, he feels much lighter than he did on Earth – his weight is only about 125 N. This is because the Moon has a smaller mass than the Earth, and attracts the astronaut with a smaller force of gravity.

His weight on Earth is 750 N.

His weight in space is 0 N.

His weight on the Moon is 125 N.

Exercises

1. Explain what is meant by the terms mass, force, force of gravity, and weight.
2. In what units are mass, force and weight measured?
3. Describe one important difference between mass and weight.

Equilibrium and centre of gravity

This section, and the rest in this chapter, deal with a subject called *statics*. This is all about balanced forces.

Equilibrium

On Earth forces are needed to move objects. But even if an object is *not* moving, there are still at least two forces acting on it. One of them is the downwards force of gravity, which acts on everything on Earth. The other force is provided by a surface which will bend until it pushes the object with an equal, upwards force which stops it from moving. An object that is not moving is *in equilibrium*. For the object to be in equilibrium, the forces must be exactly the same size, and exactly in line.

There are three types of equilibrium, depending on what happens when the forces go out of line. They can be shown by a ball bearing placed on a piece of flexible track.

Stable equilibrium. If the ball bearing is moved it returns to its original position:

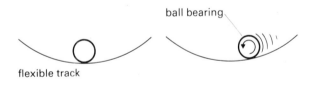

Unstable equilibrium. If the ball bearing is moved slightly it continues to move:

Neutral equilibrium. If the ball bearing is moved slightly it stays in its new position:

Centre of gravity

There is a point inside or near every object, called its *centre of gravity*. This is the point from which all the weight of the object may be considered to act. The centre of gravity of this book is in its middle, at the point where lines from opposite corners cross. The book can be balanced by supporting it with one upwards force at this point. The centre of gravity need not be in a solid part of the object as shown by the horse shoe below:

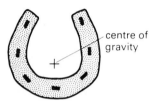

Finding the centre of gravity of a lamina. A thin sheet of material called a *lamina,* has three holes drilled near its edge. It is suspended on a pin from one hole, and a plumb line is hung in front of it:

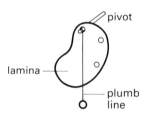

The centre of gravity falls as low as possible by lying on a line directly underneath the suspension point. The plumb line marks this line, which is marked on the lamina. The lamina is then suspended from each of the other two holes in turn, and the position of the plumb line marked again. The centre of gravity must lie on the point where the lines cross:

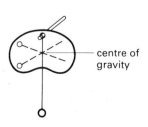

Stability of tilted objects

As long as the centre of gravity of an object lies somewhere above its base area, it remains stable. Even when the object is tilted, as long as the vertical from the centre of gravity passes through its base, its weight will pull it back:

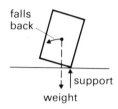

If the object is pushed so far that the vertical from the centre of gravity lies outside the base, its weight pulls it over:

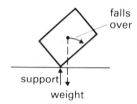

The position of the centre of gravity affects stability. A racing car is stable because its centre of gravity is low. The car has to be tilted through a large angle before the vertical from its centre of gravity lies outside the wheel base, and the car overturns. The centre of gravity of a double decker bus is higher than that of a racing car. Even so, a bus can be tilted to an angle of over 30° before it falls over:

"No standing upstairs" Why?

The famous "leaning tower of Pisa". Would it be as famous if it were straight?

Exercises

Copy the sentences 1 to 5 and complete them:

1. An object is in equilibrium when . . . .

2. An object is in stable equilibrium, if when it is moved slightly it . . . .

3. An object is in unstable equilibrium, if when . . . .

4. An object is in neutral equilibrium, if when . . . .

5. The centre of gravity of an object is . . . .

6. The photograph shows three bunsen burners in positions of equilibrium. Explain, giving a reason, which type of equilibrium each bunsen is in.

7. Give one reason why the famous leaning tower of Pisa in Italy, shown above, does not fall down – even though it leans to one side.

Moments and levers

A turning force is needed to open a gate, to turn a spanner, or to steer a car. It is the combined effect of the force and its distance from the turning point that determines the strength of the turning force:

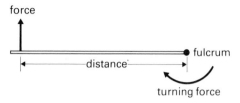

A turning force is called a *moment*. To make a turning force, a turning point such as a hinge or a pivot is always needed. Such a point is called a *fulcrum*:

moment of a force = force × distance from fulcrum

The principle of moments

A moment can act on an object in either a clockwise or an anticlockwise direction. For the object not to turn, the anticlockwise moment must be exactly equal to the clockwise moment – this is the principle of moments. A metre rule makes a good object to demonstrate this principle, and weights provide simple forces. In the next diagram, the left hand weight makes an anticlockwise moment, and the right hand weight a clockwise moment:

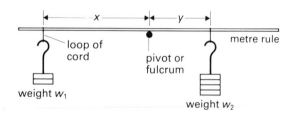

The position of the weights are altered until the rule balances. The heavier weight is seen to be closer to the fulcrum than the lighter one.

As long as the rule balances, it is found that the anticlockwise moment $w_1 \times x$ is equal to the clockwise moment $w_2 \times y$. This demonstrates the principle of moments:

When a body is in equilibrium the anticlockwise moment is equal to the clockwise moment.

Weighing by moments

The principle of moments can be used to find the weight of an unknown object. The unknown weight and a known weight are hung from a metre rule as shown below:

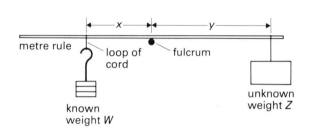

The two weights are moved about until the rule balances. The clockwise moment (unknown weight $\times y$) is equal to the anticlockwise moment (known weight $\times x$).

$$\text{unknown weight} \times y = w \times x$$
$$\text{unknown weight} = \frac{w \times x}{y}$$

The same idea is used by some balances. The scale pan is fixed on one end of a beam, and the object to be weighed is placed on it. A sliding weight is moved along the other side of the beam until it balances. The beam is marked with a scale so that the weight of the object can be read off directly:

The position of the sliding weight indicates the weight of the object.

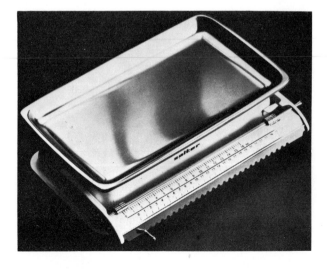

Levers

The principle of moments suggests that a small force can be used to balance a large force provided that the small force is much further from the fulcrum than the large one. A lever uses this idea to move a large load a small distance. In the example below a man exerting a force of 100 N downwards can balance a load of 1000 N. Increasing the force slightly above this amount would make him able to lift it:

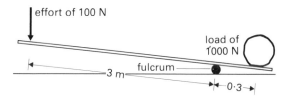

anticlockwise moment = $100 \times 3 = 300$ N.m

clockwise moment = $1000 \times 0.3 = 300$ N.m

Types of levers. Levers are used frequently in many common household tools. Three examples of the use of levers are shown here. In each case the load, effort, and fulcrum are all in different positions relative to one another:

Looking from the left – the order is: fulcrum, load, effort. Why is it best to load a wheelbarrow mostly at the front?

Looking from the left – the order in scissors and shears is: effort, fulcrum, load. If you are cutting something tough, what part of the blades will you use?

In tweezers and tongs, the order is: fulcrum, effort, load.

Exercises

1. What is meant by the moment of a force?

2. State the principle of moments.

3. Use the principle of moments to work out the unknown weight:

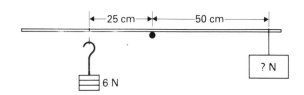

4. Copy the diagram of the two lever systems below into your book and mark on them the positions of the load, effort, and fulcrum.

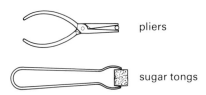

pliers

sugar tongs

Elasticity

The car springs on the right hand side of the photograph have been squashed. They will return to their original length when the car comes out of the corner.

A spring is something whose length changes when the force on it changes. It also returns to its original length when the force has been removed. It is *elastic*.

Forces stretch springs
Springs can be made which stretch when they are pulled. The bigger the force on the spring the longer it gets:

The force is provided by the weight of standard masses which are hung on the spring. The spring gets longer by exactly the same amount for each extra newton of force added. The scientific statement of this is:

The extension of a spring is directly proportional to the force on it.

The spring balance
The fact that a spring gets longer when increasing forces are placed on it is used in the spring balance. The diagram below shows an outside view and a cross-section of an extension spring balance:

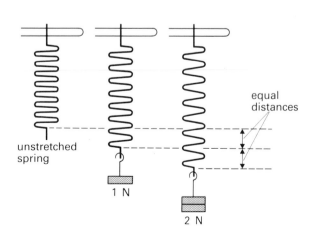

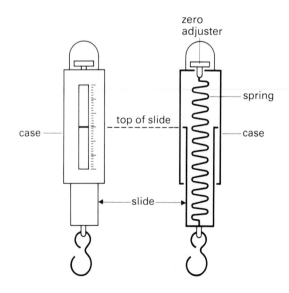

The top of the slide is visible through a slot in the balance case – it is used as a pointer. The extension of the spring, and thus the position of the pointer is proportional to the weight on the spring. Therefore a scale in newtons can be marked on the case.

A compression spring can also be used in a balance since the amount that the spring squashes depends on the force on it. A compression spring balance is shown below. The amount that the spring is compressed is recorded by the pointer at the front:

The scale measures the number of newtons of force pushing the pan downwards.

Hooke's law

The seventeenth century scientist Robert Hooke was the first person to realize that the extension of a spring keeps in step with the force on it. So this result is now called Hooke's law. A simple way of describing this is by means of a graph. When the extension is directly proportional to the force, the graph is a straight line through the origin:

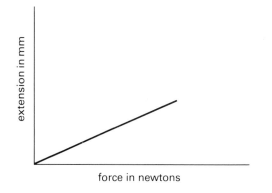

The elastic limit

If too large a force is applied to the spring two things happen.

1. The length of the spring increases by a far greater amount for each extra newton of force added. Hooke's law is no longer obeyed – the line on the graph curves upwards:

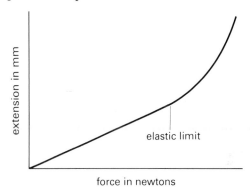

2. The spring becomes permanently stretched as shown and will no longer go back to its original length:

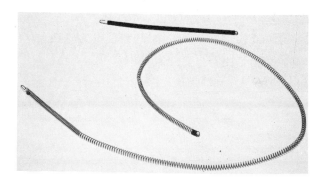

The spring has been stretched beyond its elastic limit. On the graph the elastic limit is the point where the line starts to curve upwards.

Exercises

1. What is meant by the word "elastic"?

2. Describe an experiment to find how the extension of a spring varies with the force on it. Draw a sketch of the graph of extension versus force that you would expect. Mark on the graph the elastic limit.

3. State Hooke's law.

4. Explain with the aid of a diagram how a spring balance works.

Density

The first section said that "large objects usually have a large mass, and small objects a small mass". This is not always true. A large volume of expanded polystyrene may have much less mass than a small volume of metal. Polystyrene is much less dense than metal:

A particular volume of polystyrene is lighter than the same volume of lead. Polystyrene has a lower *density* than lead. In order to compare the density of different substances it is necessary to use the same volume each time. The usual volume chosen is 1 cubic metre:

Density is defined as the mass per 1 cubic metre of a substance.

The density of any size lump of a substance is found by dividing its mass by its volume:

$$\text{density} = \frac{\text{mass}}{\text{volume}} \quad \text{kilograms/cubic metre}$$

$$d = \frac{m}{v} \quad \text{kg/m}^3$$

Density

If pieces of polystyrene and lead *of the same size* are placed on an electronic balance, the result shown below might be obtained:

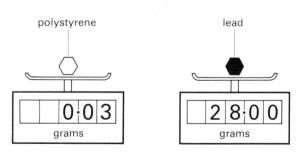

Measurement of density
To find the density of an object:

1. Measure its mass.
2. Measure its volume. Then:
3. $\text{density} = \dfrac{\text{mass}}{\text{volume}}$

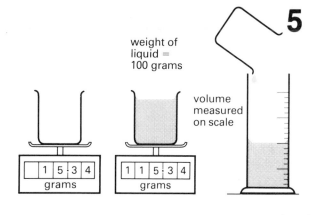

weight of
liquid =
100 grams

volume
measured
on scale

| 1 | 5 | 3 | 4 |
grams

| 1 | 1 | 5 | 3 | 4 |
grams

The following three sections deal with how to find the densities of different shaped solid objects, and of liquids. In the laboratory, grams and cubic centimetres are used rather than kilograms and cubic metres.

Density of rectangular objects
Finding the density of a rectangular object is very easy:
1. The mass is found using a balance marked to read directly in grams.
2. The volume is found by measuring the length, breadth, and height with a rule, and multiplying all three together. Then:

3. $\text{density} = \dfrac{\text{measured mass}}{\text{measured volume}}$

Density of irregular objects
In this case measuring the volume is not so easy:
1. The mass is found using a balance.
2. The volume cannot be measured directly – so it is found by measuring the volume of water that the object displaces. This is done using a displacement vessel:

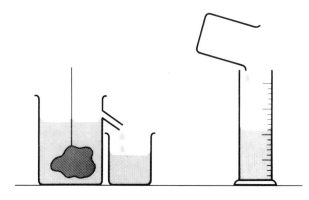

3. $\text{density} = \dfrac{\text{measured mass}}{\text{measured volume}}$

Density of a liquid
Here, finding the mass is more difficult:
1. Liquids must be held in a container. So the mass of the container must be subtracted from the mass of the liquid and container together.
2. The volume is found by pouring the liquid into a measuring cylinder. Then:

3. $\text{density} = \dfrac{\text{measured mass}}{\text{measured volume}}$

The density of water
The density of water is exactly 1000 kg/m³. The actual size of the kilogram was chosen so that this should be the case.

Relative density
As water is such a common substance, its density is taken as a reference point. The relative density of a substance shows how many times it is more or less dense than water. Iron is 7.86 times more dense than water – it has a relative density of 7.86:

$\text{relative density (R.D.)} = \dfrac{\text{density of substance}}{\text{density of water}}$

In practice relative densities are found by comparing masses of the same volume:

$\text{R.D.} = \dfrac{\text{mass of a volume of the substance}}{\text{mass of the same volume of water}}$

Relative density has no units. It is just a ratio which compares the density of a substance with that of water.

Exercises
1. Write down the formula for density.
2. In what units is density usually measured?
3. Work out the densities of the objects shown in the diagram, in grams per cubic centimetre.

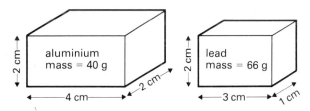

aluminium
mass = 40 g
2 cm
4 cm
2 cm

lead
mass = 66 g
2 cm
3 cm
1 cm

4. Describe an experiment to find the density of an irregular object.
5. Describe an experiment to find the density of methylated spirit.
6. Explain how you would measure the volume of a roughly shaped piece of cork.

Archimedes' principle

Upthrust reduces weight

If an object is totally or partially immersed in a liquid as shown it receives an upward force from the liquid:

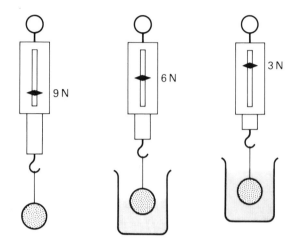

This upward force, which is called the upthrust, makes it loose weight. The loss in weight is equal to the upthrust.

The effects of upthrust can be very important. The boulders in the photograph below had lost some of their weight in the flood water of the river and so were carried down causing much damage. The upthrust does not depend upon the weight of the object, but upon the weight of liquid that it pushes out of its way, or *displaces*. The size of the upthrust is explained by *Archimedes' Principle*.

Archimedes' principle

The principle may be demonstrated using an object on a spring balance, and a displacement vessel containing water. As the object is lowered into the water, its weight gradually decreases. The water it displaces is collected. When the object is completely submerged, its loss of weight is found to be exactly equal to the weight of the water that it has displaced:

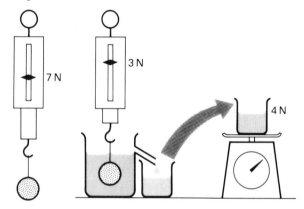

This demonstrates Archimedes' principle for water. The principle is still true even if the object is only partly immersed, or if other liquids or even gases are used instead of water. Liquids and gases are given the common name *fluids:* a complete statement of Archimedes' principle is that:

When an object is totally or partially immersed in a fluid it experiences an upthrust which is equal to the weight of fluid displaced.

Archimedes' principle in operation: Lynmouth floods, 1952.

Relative density by Archimedes' principle

It is possible to measure relative density using Archimedes' principle. The method for solids is slightly different to that for liquids.

Solids

This method works by comparing the weight of an object with the upthrust on it when it is immersed in water. The relative density of the object is high when its weight is large compared with the upthrust on it. The procedure is as follows:

1. Weigh the object;

2. Weigh the object in water, as shown in the diagram below:

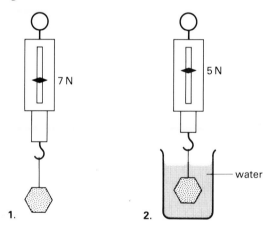

3. Work out the water's upthrust:

$$\frac{\text{water}}{\text{upthrust}} = \frac{\text{weight}}{\text{in air}} - \frac{\text{weight}}{\text{in water}}$$

4. Use the formula:

$$\text{relative density} = \frac{\text{wt. in air}}{\text{water upthrust}}$$

Liquids

This method uses any object, and compares the upthrust on it made by water, with the upthrust made by the liquid. The method is very similar to that for solids, but two more stages are added at the end:

1. Weigh any object in air;

2. Weigh the object in water.

3. Work out the water's upthrust:

$$\frac{\text{water}}{\text{upthrust}} = \frac{\text{weight}}{\text{in air}} - \frac{\text{weight}}{\text{in water}}$$

4. Weigh the object in the liquid, as in stage **3** in the next diagram:

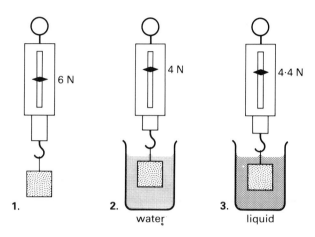

5. Work out the liquid's upthrust:

$$\frac{\text{liquid}}{\text{upthrust}} = \frac{\text{weight}}{\text{in air}} - \frac{\text{weight}}{\text{in liquid}}$$

6. Use the formula:

$$\text{relative density} = \frac{\text{liquid upthrust}}{\text{water upthrust}}$$

Exercises

1. Explain how it is that massive boulders can be carried down by the flood water of a river.

2. State Archimedes' principle and describe an experiment to demonstate it.

3. Describe an experiment to determine the relative density of a solid using Archimedes' principle. Using the figures marked as stages 1 and 2 of the last two diagrams, work out the relative density of the solids.

4. Describe an experiment to determine the relative density of a liquid using Archimedes' principle. Using the figures marked as stages **1, 2** and **3** in the last diagram, work out the relative density of the liquid.

5. An object weighs 5.6 N in air, 4.8 N in water and 4.6 N when immersed in a liquid. Calculate the relative density of the object and the liquid.

Flotation

Some things float, and some things don't. It is difficult to predict whether an object will sink or float in a liquid. A solid object will sink if its density is greater than that of the liquid, but if its shape is altered, it can be made to float.

The ability of an object to float depends on whether the liquid can provide enough upthrust to support all the object's weight. As the object sinks deeper, the upthrust on it becomes larger. If the upthrust becomes equal to the weight of the object before it is completely immersed, it floats. If the object is totally immersed and the upthrust is still not equal to its weight, the object sinks.

The principle of flotation
This states that:

A floating body displaces its own weight of fluid.

This can be demonstrated using the apparatus shown below. The weight of an object that can float is compared with the weight of water it displaces:

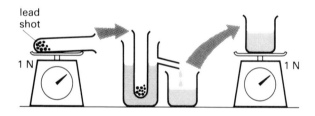

When the weight of the tube is compared with the weight of water it displaces, the two are found to be the same.

Floating in different liquids
If a particular object is floated in liquids of different density, it is found that it floats deeper in the less dense liquids. More liquid has to be displaced before the object can displace its own weight of liquid:

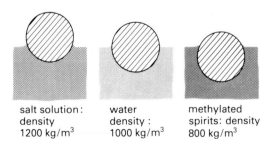

salt solution: density 1200 kg/m³

water density: 1000 kg/m³

methylated spirits: density 800 kg/m³

Fresh water is less dense than sea water. This means that ships loaded whilst in sea water may sink dangerously deep when they sail into fresh water. To prevent this happening a *Plimsol Line* is marked on the side of the ship, which shows the maximum loading level when the ship is in different types of water:

Hydrometers
The same effect is also used in an instrument called a *hydrometer*. A hydrometer is a special type of float, which is used to measure the densities of liquids. It sinks deeper into liquids of lower density. The actual density is read off from the position of the liquid surface against a scale marked off on the side, as shown in the diagram at the top of the next page.

A car battery hydrometer is a special type of hydrometer which indicates how well charged a car battery is. It does this because the car battery contains dilute sulphuric acid which becomes more concentrated, and thus more dense, as the battery is charged. This causes the hydrometer to float higher. The rubber tube is placed in the acid of the battery, and the bulb is squeezed to expel the air from it. When the bulb is released, suffcient acid rises to float the hydrometer so that the density can be read off.

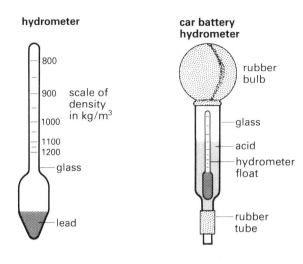

hydrometer

— 800

— 900 scale of
         density
— 1000 in kg/m³

— 1100
— 1200

— glass

— lead

car battery
hydrometer

— rubber
   bulb

— glass

— acid

— hydrometer
   float

— rubber
   tube

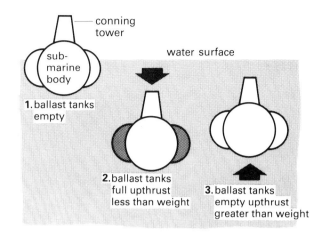

— conning
   tower

water surface

sub-
marine
body

1. ballast tanks
   empty

2. ballast tanks
   full upthrust
   less than weight

3. ballast tanks
   empty upthrust
   greater than weight

The submarine

The submarine makes use of the principle of flotation. It contains special tanks called ballast tanks, which can contain either air or water. With air-filled ballast tanks, the upthrust on the half submerged submarine is equal to its weight, so the submarine floats (diagram **1**). But if the ballast tanks are allowed to fill with water, the submarine becomes heavier. The upthrust is no longer sufficient to support the weight of the submarine so it sinks (diagram **2**).

The submarine carries cylinders of compressed air. The air is used to force the water out of the ballast tanks. The weight of the submarine is now less than the upthrust, so the submarine rises (diagram **3**):

The ballast tanks can be partly filled so that the weight of the submerged submarine is exactly equal to the upthrust. The submarine then 'hangs' in the water, like a fish. It can be driven at an angle to the surface using the engines and the horizontal fins shown in the photograph below.

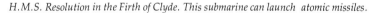

Exercises

1. State the principle of flotation and describe an experiment that demonstrates it.
2. Draw diagrams showing a hydrometer floating in a liquid of density **a** 800 kg/m³ and **b** 1200 kg/m³.
3. Draw a diagram of a car battery hydrometer and explain how it is used.
4. What is the purpose of a ship's Plimsol Line?
5. How is a submarine able to dive and surface?

H.M.S. Resolution in the Firth of Clyde. This submarine can launch atomic missiles.

Questions on chapter 1

1. Copy the diagrams of the following objects into your book. Mark with a cross the approximate position of the centre of gravity in each case.

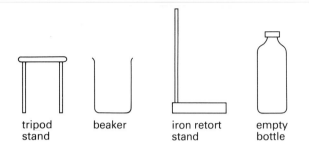

tripod stand   beaker   iron retort stand   empty bottle

Write down the name of the most stable object shown and explain why it is the most stable.

(Y.R.E.B.)

2. Copy into your book the diagram of the crowbar being used to lift a heavy stone.

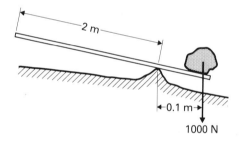

i Mark with an arrow on your diagram where you would push, and in what direction, to lift the stone with the *least* effort.
ii Calculate the force which should be applied in order to just lift the stone i.e. just to balance the weight of the stone.   (M.R.E.B.)

3. The diagram shows a see-saw with a weight of 200 N at the end.

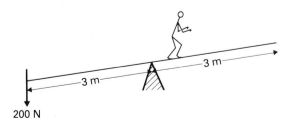

A person weighing 600 N moves from the pivot away from the 200 N. How far will this person have to move to balance the see-saw?   (E.A.E.B.)

4. A student hung weights on the end of a spring and measured the extensions produced. He then plotted the graph of load against extension shown below, and claimed that this spring obeyed Hooke's law.

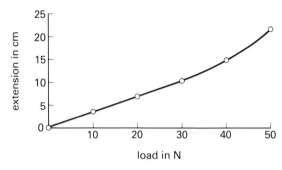

i State Hooke's law as applied to springs.
ii Explain whether he was correct or not in claiming that his spring obeyed Hooke's law.
iii Estimate the extension of this spring for a load of 25 N.
iv Estimate the elastic limit of this spring in newtons.   (W.M.E.B.)

5a Look at the diagram below.

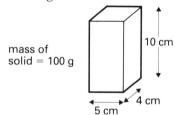

mass of solid = 100 g

10 cm
4 cm
5 cm

i Calculate the volume of the solid.
ii Calculate its density.
iii Would it float if placed in water of density of 1 g cm^{-3}?
iv When placed in water what volume of water would be displaced?
v What fraction of the volume of the solid would be immersed?

b Another solid of the same shape and size has a mass of 300g.

i Calculate its density.
ii Would this solid float if placed in water?
            (M.R.E.B.)

6. The diagram shows a 1 kg metal block, **M**, hung on a spring balance. The reading on the balance was 10 N. The block was then lowered into the water in a displacement can so that **M** was completely immersed, and the overflow was collected in a measuring cylinder. The spring balance read 6 N.

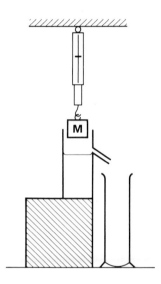

i What is the decrease in reading on the spring balance?
ii What weight of water, in newtons, had overflowed into the measuring cylinder?
iii If the weight of a 100 g mass is 1 N, what is the mass of water displaced?
iv If 1 g of water has a volume of 1 cm³, what is the volume of water in the measuring cylinder?
v From the results obtained find the density of the metal. (A.L.S.E.B.)

7. State Archimedes' principle.
A diving bell, of weight 60 000 N and volume 2 m³, is to be raised from the bottom of the sea. If the density of sea water is 1024 kgm⁻³ calculate:
  i The mass of sea water displaced by the bell.
  ii The weight of sea water displaced by the bell, given that a mass of 1 kg has a weight of 10 N,
  iii The upthrust in newtons,
  iv The force a crane must exert to just lift the bell from the sea bed. (W.J.E.C.)

8. The diagram shows the same instrument floating in three different liquids: paraffin, water, salt water.

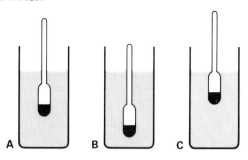

i Copy the diagram into your book and under each one write the name of the liquid.
ii What is the instrument called?
iii State one everyday use of this instrument. (E.A.E.B.)

9. The diagram shows an instrument being used to measure the density of a liquid.

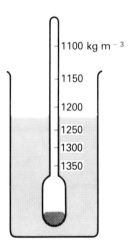

i Name the instrument.
ii Write down the density of the liquid which is shown by the instrument.
iii What causes the instrument to float upright?
iv If the instrument were put into water of density 1000 kg m⁻³, what would happen?
v If one of these instruments was not available and you needed to measure the density of the liquid, list the apparatus you would need and describe how it would be used. Show how the density would be calculated from these results. (Y.R.E.B.)

21

Speed, velocity, and acceleration

Three different terms are used to describe the rate at which things move from one place to another: speed, velocity and acceleration.

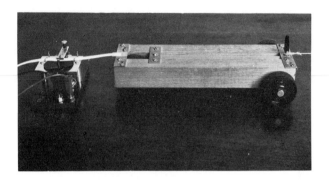

Speed

Speed is a measure of how fast things are moving. A car travelling at high speed moves a greater distance in the same time than a car travelling at slow speed. Speed can be measured by finding what distance is travelled in a given time. In other words:

Speed is a measure of the change of distance as time passes.

$$\text{speed} = \frac{\text{distance covered}}{\text{time taken}}$$

In the laboratory, speeds are measured in metres per second (m/s).

Velocity

The word *velocity* is often used instead of speed, but velocity means more than just speed. A car can speed either away from, or towards, a certain place. The *velocity* of the car indicates how fast it is moving, and also the *direction* in which it is moving:

Velocity is speed in a particular direction.

A car travelling from **A** to **B** with a speed of 50 m/s has a velocity of 50 m/s. But the car travelling in the opposite direction from **B** to **A** has a velocity of −50 m/s.

Acceleration

When a car speeds up it is said to be *accelerating*. The amount of distance it is able to cover in each second is gradually increasing:

Acceleration is a measure of the change of speed as time passes.

$$\text{acceleration} = \frac{\text{speed change}}{\text{time for speed to change}}$$

A car that increases its speed from 0 m/s to 10 m/s in one second is accelerating at 10 m/s/s, or 10 m/s² (ten metres per second squared). If the car slows down, it is said to be *decelerating*.

The dynamics trolley

This is a useful piece of apparatus for demonstrating speed and acceleration. The wheeled trolley pulls a paper tape behind it. A fixed vibrator makes a mark on the tape every fiftieth of a second. If the trolley moves slowly, the dots are close together. If the trolley moves fast, the dots are further apart, since the trolley moves a larger distance each fiftieth of a second:

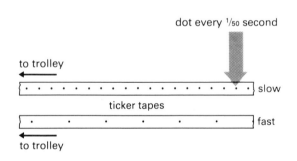

If the trolley is accelerating, the dots get further and further apart. The distance travelled each time between marks being made is increasing. If it is decelerating, they get closer and closer together:

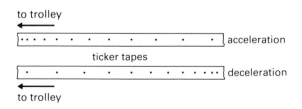

Speed of the trolley. This may be found by measuring the distance it travels in a given time. The first dot on the tape marks zero time. After five more dots $^1/_{10}$ second has passed. The distance between the first and last dot is the distance travelled in $^1/_{10}$ second.

$$\text{speed} = \frac{\text{distance change}}{\text{time for distance change}} \text{ metres/second}$$

$$v = \frac{s}{t} \text{ m/s}$$

Acceleration of the trolley. The acceleration of the trolley may be found by applying:

$$\text{acceleration} = \frac{\text{speed change}}{\text{time for speed change}}$$

$$= \frac{\text{final speed} - \text{original speed}}{\text{time for speed change}}$$

$$a = \frac{v - u}{t} \text{ m/s}^2$$

When the car is decelerating, the final speed is *less* than the original speed. This makes the acceleration negative – deceleration is "negative acceleration"!

The acceleration due to gravity

The earth exerts a gravity force on all objects near it. An object that is supported will not move, but if the supporting force is removed, the object will instantly begin to fall, accelerating as it does so. The acceleration depends on the mass of the Earth and not the mass of the object. This means that if a brick and a marble are dropped out of a window together they will both hit the ground at the same time. This acceleration is called the *acceleration due to gravity*. It is an important quantity and it is given the symbol g. The value of g for the earth is about 10 m/s².

Experiment to measure g. The acceleration due to gravity may be found by accurately measuring the length of time it takes for a ball bearing to fall a measured distance. The apparatus used is shown at the top of the next column.

When the double switch is opened the electromagnet is switched off and the timer starts. The ball bearing falls and knocks open the bottom switch which stops the timer. In this way the time t taken for the ball to fall a distance s is measured.

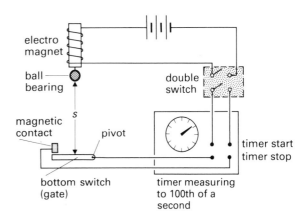

The acceleration due to gravity is then worked out from the formula:

$$\text{acceleration} = \frac{2 \times \text{distance}}{(\text{time})^2} \text{ metres/second}^2$$

$$g = \frac{2s}{t^2} \qquad \text{metres/second}^2$$

Measuring wells. The same formula can be turned round and used to find the depth of a well. A stone is dropped down a well. The time t between releasing the stone and hearing the noise of it hitting the bottom is measured. Then the depth of the well s is given by:

$$\text{depth} = \frac{\text{gravity acceleration} \times (\text{time})^2}{2}$$

$$s = \frac{g\,t^2}{2} \text{ metres}$$

Since $g = 10$ m/s² the formula can be simplified to depth of well = $5t^2$ metres.

Exercises

1. Explain what is meant by the terms speed, velocity, and acceleration.
2. Write down the formulae for speed and acceleration stating what the symbols mean.
3. A motor bike travels a measured 100 metres in 4 seconds. What is its speed in metres per second?
4. A car's speed increases from 10 m/s to 30 m/s in 2 seconds. What is its acceleration?
5. Describe an experiment to measure the acceleration due to gravity.
6. A boy drops a stone down a well and hears the sound of it hitting the bottom 2 seconds later. How deep is the well?

Graphs of motion

A graph is a chart that shows how one thing varies in response to changes in something else. For example, a graph can be drawn that shows how the distance of a car from its starting point changes as time passes. This is a *distance-time* graph. A graph showing how the car's velocity changes as time passes is a *velocity-time* graph.

Distance time graphs

If an object is stationary its distance from a particular point does not change as time passes. Each second, it is the same distance away. This is shown on a *distance-time* graph by a horizontal line:

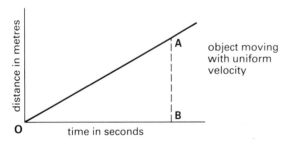

An object moving with steady or uniform velocity travels equal distances in equal times. The object's distance from a point increases by an equal amount each second. This is represented on a *distance-time* graph by a straight line at an angle:

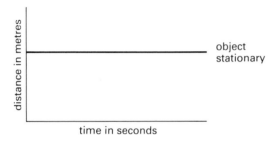

The velocity of the object is found by measuring the slope or *gradient* of the line:

$$\text{velocity} = \frac{\text{distance change}}{\text{time for distance change}}$$
$$= \text{gradient of } \mathbf{OA}$$
$$= \frac{\mathbf{AB}}{\mathbf{OB}} \text{ metres/second}$$

For an object whose velocity is increasing, the distance travelled in each second increases, so the slope of the graph increases. In other words, its gradient increases – the graph curves upwards. For an object whose velocity is decreasing the gradient

gets less as the graph bends over towards the horizontal:

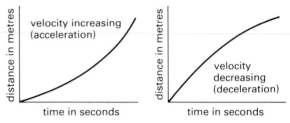

Velocity-time graphs

Graphs to show how the *velocity* changes with time can also be drawn. A horizontal line now represents a steady *velocity*, not a steady distance:

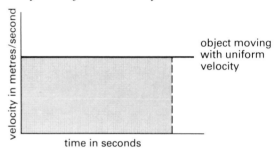

The distance travelled by the object can still be seen on the graph. Remember that:

distance = velocity × time
= shaded area in graph

The distance travelled in a certain time is given by the area under the graph up to that time.

Steady accelerations and decelerations (changes in velocity as time passes) are represented by sloping lines as shown below:

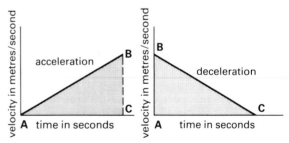

The size of the acceleration is found by measuring the gradient of the line:

$$\text{acceleration} = \frac{\text{velocity change}}{\text{time for velocity change}}$$
$$= \text{gradient of } \mathbf{AB}$$
$$= \frac{\mathbf{BC}}{\mathbf{AC}} \text{ metres/second}^2$$

Can you draw the distance-time graph for this vehicle?

Example. The graph below shows how the velocity of a motor bike varies over a period of 50 seconds. It accelerates steadily for 10 seconds up to a velocity of 20 m/s. It continues with uniform velocity for a further 20 seconds and then decelerates so that it stops in 20 seconds. Work out **1** the acceleration, **2** the deceleration, and **3** the distance travelled.

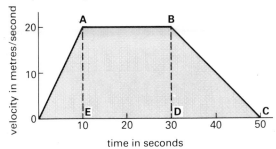

1. acceleration = $\dfrac{\text{velocity change}}{\text{time for velocity change}}$

= gradient of **OA**

= **AE/OE** = 20/10

Acceleration = 2 m/s²

2. deceleration = $\dfrac{\text{velocity change}}{\text{time for velocity change}}$

= gradient of **BC**

= **BD/DC** = 20/20

Deceleration = 1 m/s²

3. distance travelled = area under the graph

= area of trapezium **OABC**

= ½ (**AB** + **OC**) × **AE**

= ½ (20 + 50) × 20

= ½ × 70 × 20

Distance travelled = 700 m

(you could also find the area by adding up the areas of the two triangles and the rectangle, if you did not remember the trapezium formula).

Exercises

1. Draw *distance-time* graphs to show (a) increasing velocity, (b) uniform velocity, and (c) decreasing velocity. Write a sentence to explain each one's shape.

2. Explain what is happening in this *distance-time* graph, for a stone thrown vertically up into the air:

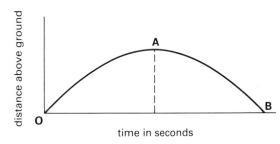

3. Draw *velocity-time* graphs to show (a) acceleration, (b) deceleration, and (c) uniform velocity. Write a sentence to explain each one's shape.

4. A brick is dropped from a window. Make a sketch of the *distance from the ground – time* graph which shows its motion.

5. A car accelerates to a velocity of 30 m/s in 10 seconds, continues at this velocity for a further 30 seconds and then decelerates for 20 seconds so that it stops. Draw a *velocity-time* graph to represent this motion and calculate (a) the acceleration, (b) the deceleration and (c) the distance travelled.

Newton's laws

Newton's laws are about forces and how they cause movement. The first law explains what a force does, the second how to measure the size of a force, and the third explains how forces produce reactions.

Newton's first law

An object will not start to move on its own. It needs a force to make it move. Once an object is moving it needs another force to stop it. A force is also needed to change the direction of a moving object:

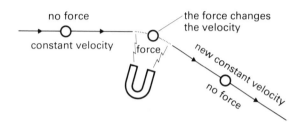

All this is summarised in Newton's first law which states:

An object remains still or moves with constant velocity unless it is acted on by a force.

The car in the photograph has been forced to stop by the wall:

Testing for effective seat belt design.

The dummy driver carries on with constant velocity until it is forced to stop by the steering wheel and the windscreen. The dummy would be more likely to remain in one piece if the necessary force to stop it had been applied by a seat belt.

Newton's second law

When there is only one overall force acting on an object, it accelerates. The strength of the force is measured by its ability to accelerate the mass. A large force will give a big mass a big acceleration. A small force will give a small mass a small acceleration. Newton's second law summarises this by saying that:

The force on an object is measured by the product of the mass of the object and its acceleration.

force = mass × acceleration
$F = ma$

The unit of force is the *newton*. The size of 1 N is specially chosen so that it gives a mass of 1 kg an acceleration of 1 m/s².

Newton's second law may be demonstrated using a dynamics trolley. A trolley of known mass is caused to accelerate by being pulled along by a known force. The force is caused by weights, made so that each one provides exactly 1 N of force:

It is found that a force of 10 N gives a trolley of mass 1 kg an acceleration of 10 m/s². Newton's second law states that:
$F = m \times a$; putting in the numbers,
$10 = 1 \times 10$ which fits in with the law.

If the mass of the trolley is doubled by stacking a second trolley on top the same force gives it an acceleration of only 5 m/s². In this case:

$F = m \times a$

$10 = 2 \times 5$ which also fits in.

Newton's third law

Forces pull things together or push things apart. The force of gravity is pulling this book and the Earth together. They will move towards one another until the ground and the book meet. Then the book and the ground will press equally on one another. The force of the book pressing down on the ground (its weight) is called an *action* and that of the ground pressing up on the book a *reaction*. All forces behave similarly, and this effect is summarised by Newton's third law which states:

Action and reaction are equal and opposite.

The space rocket works because the force produced by burning the fuel pushes the engine and hot gases apart. For convenience the force is considered to push the hot gases out of the engine. This is the *action*. The *reaction* then pushes on the rocket. The rocket can "lift off" only when the reaction force is greater than its weight.

Newton's third law can be demonstrated for stationary forces by using two spring balances set up as shown in the next diagram:

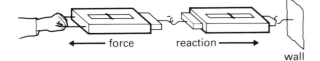

← force   reaction →   wall

The left-hand balance is pulled with a force and it is found that the balance fixed to the wall registers the same force. The wall is pulling back with an equal reaction.

Newton's third law is shown for moving forces by a dynamics trolley. It has a catapult mounted on it:

The catapult is released by burning the string with a taper. The tension in the elastic sends the ball bearing backwards. The reaction to this force pushes the trolley forwards.

Exercises
1. State Newton's three laws.
2. Write down the equation for Newton's second law. State what the symbols mean and what units they are measured in.
3. A car has a mass of 1000 kg and its engine will provide a driving force of 5000 N. What is the acceleration?
4. What force is needed to give a mass of 10 kg an acceleration of 65.2 m/s²?
5. A force of 30 N gives an object an acceleration of 7.5 m/s². What is the mass of the object?

Work, power, and energy

Work, energy and power are words which have particular meanings in Physics.

Work

When a force moves an object through a distance it does *work*. It does not matter whether the force is accelerating the object (such as a rocket taking off) or is simply moving it along (such as pushing a car that has run out of petrol). The work that has been done is found by multiplying the force by the distance it has moved:

work = force × distance force moves

$$W = Fs$$

Work is measured in *joules* (J). The value of the joule is specially chosen so that 1 J of work is done when a force of 1 N moves a distance of 1 m.

Power

Power is used to measure the rate at which work is done. Both a powerful man and a weak man are capable of doing a manual job such as lifting a large number of heavy weights from a low shelf to a high one. But the powerful man can do it in a shorter time than the weak man – he has more *power*. Power is measured by dividing the work done by the time taken to do it:

$$\text{power} = \frac{\text{work done}}{\text{time taken}} \text{ watts}$$

$$P = \frac{W}{t} \text{ W}$$

Power is measured in watts (W). The value of the watt is specially chosen so that a power of 1 W means that 1 J of work can be done in 1 s.

A pupil running up stairs as fast as he can has a power of about 500 W, which is about 1½ times as powerful as an electric drill.

Energy

A person with a lot of energy is able to do a lot of work. He may wish to do it fast or slow, but he is capable of doing a lot. Energy may be thought of as "stored work". Put scientifically:

Energy is the capacity for doing work.

Energy, like work, is measured in joules. There are various forms of energy, as shown in the next diagram:

type of energy	explanation
kinetic	Energy stored in the movement of an object, released by stopping it. kinetic energy = ½ mass × (velocity)^2 $= \frac{1}{2}mv^2$
potential	Energy stored by virtue of an object's position above the ground. When the objects falls the energy is released: energy = mass × $\dfrac{\text{gravity}}{\text{acceleration}}$ × height $= mgh$ **or** Energy stored by virtue of an objects shape e.g. a coiled spring.
chemical	Energy locked up in a chemical compound. It is released in a chemical reaction such as burning.
heat	Energy of hotness. Heat energy is difficult to store, because it "leaks" away to cooler places.
light	Energy that can be seen. Light energy is impossible to store because it radiates away, or turns to heat.
sound	Vibrational energy. This energy cannot be stored either, but travels away from the sound source through the surrounding material.
nuclear	Energy stored in the structure of the atom. This energy is released in a nuclear reaction as in the atomic bomb.

Energy changes

Mass and energy can be converted from one to the other. One type of energy can also be converted into another type of energy. This is summarised in the law of conservation of mass-energy which states:

Electrical wind generator

Installing electricity cables

*Energy changes
in a test motor car*

Mass and energy can neither be created nor destroyed but they can be converted from one form to the other.

The following examples show how energy is converted from one form to another.

Kinetic energy to electrical energy. The kinetic energy of the blowing wind turns the blades of the windmill. The blades drive an electric generator. In this way some of the kinetic energy of the wind is converted into electrical energy.

Potential energy to kinetic energy to heat and sound. The maintenance man in the photograph has increased his potential energy. Should he fall, he will accelerate towards the ground and his potential energy will be converted to kinetic energy. When he hits the ground his kinetic energy is converted into heat and sound, and unfortunately work will be done on his body in the process.

Nuclear energy to heat to kinetic energy to electrical energy. In a nuclear power station, energy stored inside uranium atoms is released as heat. The heat energy is used to make steam which drives the fan blades inside a steam turbine round. This movement drives an electrical generator which converts the kinetic energy into electrical energy.

Exercises

1. Write down the formulae for work and power. State what the symbols mean and the units they are measured in.

2. A crane lifts a weight of 100 N through 25 m. Calculate the work done by the crane. If the lifting takes 5 seconds what is the power of the crane?

3. A pupil of weight 600 N climbs a flight of stairs 3 m high in 4 seconds. What work has he done in lifting his own weight and what is his power?

4. Copy out the following passage, filling in the missing words from the lists.

When * is burnt in the engine of a motor bike * energy is released. The * converts the * energy into the * energy of movement of the motor bike.

(engine, chemical, petrol, kinetic, stored)

5. Describe, as accurately as you can the series of energy changes shown in the series of car photographs above.

Friction

It is difficult to pull a sledge along a dry pavement. The force that is resisting the motion of the sledge, is called *friction*, and it is caused by the sledge runners rubbing on the pavement. If the pavement is covered with snow the sledge can be pulled more easily because the friction is less.

Water resists the motion of boats, and air resists the motion of aircraft. This is fluid friction, which is called *viscosity*.

Friction in solids

Friction can be investigated in more detail using the apparatus shown in the diagram:

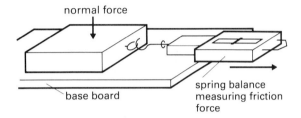

normal force

base board

spring balance measuring friction force

The block of wood is pulled by a spring balance which measures the force pulling it. The force pressing the block against the base board is called the *normal force*. In this case the normal force is caused by the weight of the block.

As the force applied through the spring balance is increased from zero, the friction force increases with it, until sliding suddenly occurs. The size of the force needed to keep it moving then drops to a lower value. The maximum value of the friction force just before sliding occurs is called the *limiting friction*. The value of the friction force during sliding is called the *dynamic or sliding friction*.

If the normal force is increased by placing weights on the block, it is found that both the limiting and dynamic friction increase in step with it.

Theory of friction

If smooth looking surfaces are examined under a high power microscrope, their actual roughness can be seen. They only touch where their high spots meet:

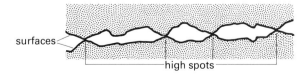

surfaces

high spots

The high spots that are touching tend to stick together. The limiting friction is the force needed to separate these high spots. Once the high spots have been separated a lower force is needed to keep the two surfaces moving.

If the normal force is increased the surfaces are squashed together more. The high spots, where the surfaces are in contact, are larger:

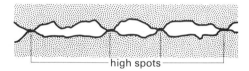

high spots

A greater limiting friction is needed to separate the high spots and a greater dynamic friction is also needed to keep the surfaces sliding.

Applications of friction. Friction can be either a nuisance that must be eliminated, or an asset that is needed.

Friction between the moving parts of a car engine is a nuisance. The heat generated by the rubbing of the parts would be so great that they would partially melt and weld themselves together. This problem is overcome by *lubrication*. As liquid friction is much less than solid friction, the rubbing surfaces are separated by a layer of oil which is pumped between them.

On the other hand friction between the road surface and the car's tyres is essential. Without friction the wheels of the car would just spin uselessly round! Car tyres are designed to have a high friction between themselves and the road, even when the road is wet:

These stunt-parachutists are in "free fall". Do you think the team uses a lot of girls?

Friction in fluids: viscosity

Water resists the motion of boats, and air resists the motion of aircraft. This is fluid friction, which is called *viscosity*. Viscosity tends to prevent motion within the fluid – it is caused by internal friction between moving layers of fluid.

The viscosity of fluids varies greatly. Water is much easier to pour than treacle. It is less viscous. Gases have a much lower viscosity than liquids.

The viscosity of all fluids gets less as they get hotter. Warm treacle for example is quite "runny".

Terminal velocity. As an object falls through a fluid, the force of gravity tries to increase its velocity. The viscous force or fluid friction, which increases as the object speeds up, acts so as to slow it down. At a certain velocity, which depends on both the viscosity of the fluid and the shape of the object, the two forces balance each other. The object now moves with a constant velocity, which is called the terminal velocity.

The sky divers in the photograph above are falling through the air with a terminal velocity of about 90 m/s. As they approach the ground, the divers open their parachutes, which greatly increases the viscous force. This reduces their terminal velocity to only about 7 m/s because of the broad shape of the parachute.

Exercises
1. What is friction?
2. Describe an experiment to show how the friction force between two surfaces depends on the normal force pressing them together.
3. How can friction be reduced?
4. Give two examples of friction being of use and friction being a nuisance.
5. What is viscosity? How does temperature affect it?
6. Oil companies used to make two grades of engine oil, a winter grade and a summer grade. Why do you think that they did this and what was the difference between the two grades? The two grades of oil have now been replaced by one which they call "viscostatic". What does viscostatic mean?
7. Explain what is meant by terminal velocity and state what two factors affect it.

Machines — pulley systems

A machine is a device that changes the size of a force and enables work to be done more easily. Machines are often used to change a small force moving through a large distance, into a larger force moving a smaller distance. This section deals with the particular example of *pulley systems,* such as the one used on the crane in the photograph. However, the terms and formulae used apply to any machine.

Pulley systems
The pulley system in the diagram below has four pulleys mounted in two blocks. The pulleys in each block are usually side by side but they are drawn below one another for clarity:

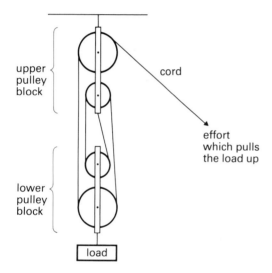

Velocity ratio. For the load to be lifted by 1 cm each of the four pieces of cord supporting the lower pulley block must be shortened by 1 cm. In order to do this the effort must pull a total distance of 4 cm. The distance moved by the effort, divided by the distance moved by the load, is called the *velocity ratio*. In this case it is 4:

$$\text{velocity ratio (V.R.)} = \frac{\text{distance moved by effort}}{\text{distance moved by load}}$$

The velocity ratio for a pulley system can be found by counting the number of strings between the pulleys.

Mechanical advantage. The load is larger than the effort – the load divided by the effort is called the *mechanical advantage*. The mechanical advantage can only be found by experiment as the force due to friction, which tries to stop the pulley wheels turning, cannot be calculated:

$$\text{mechanical advantage} = \frac{\text{load}}{\text{effort}}$$

Efficiency of a machine
In a machine with no friction, a small force moving a large distance would be exactly converted to a large force moving a small distance. The work got out would be exactly equal to the work put in. The machine would be totally *efficient*.

This is generally not the case because some of the force of the effort must be used to overcome the friction of the moving parts, and also to lift parts of the machine such as the lower pulley block. The efficiency of the machine compares the useful work got out with the work put in.

$$\text{efficiency} = \frac{\text{useful work got out}}{\text{work put in}}$$

As the useful work got out is less than the work put in, the efficiency is always less than 1.

For convenience, the efficiency of a machine is worked out from the formula:

$$\text{efficiency} = \frac{\text{mechanical advantage}}{\text{velocity ratio}}$$

Both formulas give the efficiency – both compare the amount of work being done by the effort, with the amount of work that reaches the load.

The efficiency of a machine is sometimes expressed as a percentage. In this case the formula is written as:

$$\text{efficiency} = \frac{\text{mechanical advantage}}{\text{velocity ratio}} \times 100\%$$

Finding the efficiency of a pulley system
The apparatus used is shown in the diagram opposite. The velocity ratio is found by counting the number of strings between the upper and lower pulley block.

The mechanical advantage is determined by finding the effort that is just needed to make the load move up with constant velocity. Then the mechanical advantage is worked out from:

$$\text{mechanical advantage} = \frac{\text{load}}{\text{effort}}$$

Here, the arm of the crane can be raised.

Here, the load hangs from a moveable trolley

The efficiency is now found by substituting in the formula:

$$\text{efficiency} = \frac{\text{mechanical advantage}}{\text{velocity ratio}}$$

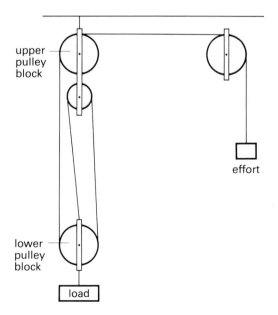

upper pulley block

lower pulley block

effort

load

If the experiment is repeated using different loads, it is found that the efficiency increases with the load. This is because the effort has to lift both the lower pulley block and the load. As the load gets bigger, the lower pulley block becomes a smaller part of the total weight lifted.

Exercises

1. Explain what is meant by the efficiency of a machine.

2. Write down the formula for velocity ratio, mechanical advantage, and effeciency.

3. Draw a diagram of a pulley system with (a) a velocity ratio of 3, and (b) with a velocity ratio of 5.

4. Work out the velocity ratio, mechanical advantage, and efficiency of the pulley system shown in the diagram on the left. The load is 4 N and the effort is 2 N.

5. The crane in the photograph is steadily lifting a load of 1 500 N. The engine is providing a force of 1 000 N. What is the efficiency of the crane?

More machines

This section describes various machines and how to work out their velocity ratios. In each case, the mechanical advantage can only be determined from an experiment which finds the smallest effort that will make a load move with steady speed. The mechanical advantage is then found from:

$$\text{mechanical advantage} = \frac{\text{load}}{\text{effort}}$$

Once these have been worked out, the efficiency of the various machines can also be found, by using:

$$\text{efficiency} = \frac{\text{mechanical advantage}}{\text{velocity ratio}}$$

The efficiency of each of the following examples can be found by using this formula.

The inclined plane

The inclined plane makes use of the fact that it is easier to push a heavy object up a slope than to lift it directly. The diagram shows the laboratory apparatus used to investigate the inclined plane:

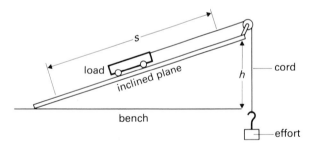

For the load to be raised through a distance h, the effort must move the truck the length of the plane s. The velocity ratio is therefore given by:

$$\text{velocity ratio} = \frac{\text{height of plane } h}{\text{length of plane } s}$$

The wheel and axle

This is investigated in the laboratory using the apparatus shown at the top of the next column. It consists of a large diameter wheel, and a small diameter axle, both of which are firmly attached to one another:

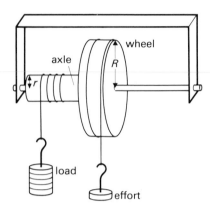

A small effort hung from a cord on a large wheel, is able to lift a large load hung from a narrow axle that has been fixed to it. In one rotation the effort unwinds by the length of the circumference of the wheel, and the load winds up by the much smaller circumference of the axle. The velocity ratio is given by:

$$\text{velocity ratio} = \frac{\text{radius of wheel } R}{\text{radius of axle } r}$$

The wheel and axle has many applications, two of which are shown below:

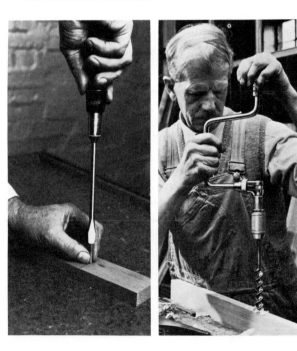

The effort in the brace and bit is applied by hand, and the load is caused by the resistance of the wood to the cutting of the bit.

The screwdriver also uses the principle of the wheel and axle. The effort is applied to the large diameter handle, and the load is caused by the resistance of the wood to the turning of the screw.

The screw thread

Diagram **1** shows a screw thread and nut. The distance between each thread is called its *pitch*. Each turn of the screw moves the nut up or down by the pitch of the thread:

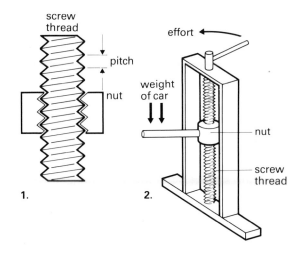

1.

2.

This idea is used on some car jacks as shown in diagram **2**. The bar is fixed to the nut and slots into the car chassis to support it. The screw thread is rotated by a lever at the top:

$$\text{velocity ratio} = \frac{2\,\pi \times \text{length of effort lever}}{\text{pitch of thread}}$$

Gears

The gears shown below do two things. First, they change the speed of a rotating shaft and second, they alter the force produced by it. The force increases as the speed decreases:

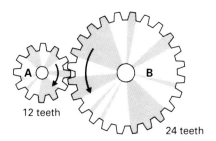

12 teeth

24 teeth

As gear **A** turns, its teeth mesh with those of gear **B** and make it turn in the opposite direction. Since it has half as many teeth, by the time **A** has gone round once, **B** has only gone round half a turn. The speeds of gears are given by the formula:

$$\text{velocity ratio} = \frac{\text{speed of gear } \mathbf{A}}{\text{speed of gear } \mathbf{B}}$$

$$= \frac{\text{number of teeth on } \mathbf{B}}{\text{number of teeth on } \mathbf{A}}$$

Note that the fastest turning gear is the one with the smallest number of teeth.

Exercises

1. Make a list of the formulae for the velocity ratios of four different machines including the pully system.

2. Copy the diagram below, using circles to represent the gears. Mark on the direction of rotation of gear **B**. Work out the speed of gear **B** and mark it on the diagram. What is the velocity ratio of the system?

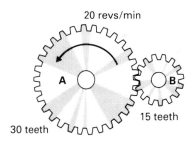

20 revs/min

30 teeth

15 teeth

3. In the wheel and axle diagram, the wheel's radius R is 10 cm, and the axle's radius r is 2 cm. The effort is 1 N and the load is 4 N. Work out the velocity ratio, the mechanical advantage, and the efficiency of the system.

4. In the inclined plane diagram, the length of the plane s is 1 m and the height h is 0.25 m. The effort is 35 N and the load is 105 N. Work out the velocity ratio, the mechanical advantage, and the efficiency of the system.

5. Describe an experiment to find the efficiency of an inclined plane. Draw a diagram of the apparatus and explain what measurements you would take. Give two reasons why the efficiency is less than 100%.

Questions on chapter 2

1. The dots show the position of a car at 10 second intervals during journeys along a particular stretch of road of length 1000 metres.

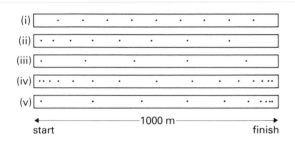

(i)
(ii)
(iii)
(iv)
(v)

← ————1000 m———— →
start   finish

a For each journey, describe in words the motion of the car. The first one is done as an example:
  i "The car moves at a constant speed over the whole 1000m."
b Taking measurements from the diagram find how far the car moved in Journey **i** in 10s.
c At what speed did the car travel in journey **i**?
d During one journey, the car increased its speed from 4 ms⁻¹ to 24 ms⁻¹ in 10s. What was its acceleration?   (S.W.E.B.)

2. Shown below are ticker tape charts illustrating three different types of motion of a body. Each tape strip in the chart shows the distance travelled by the body in equal time intervals.

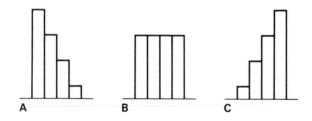

A   B   C

Describe the motion of the body shown by
  i Chart **A**
  ii Chart **B**
  iii Chart **C**

Give one example of a body moving in such a way as to produce
  iv Chart **A**
  v Chart **B**
  vi Chart **C**   (E.M.R.E.B.)

3.

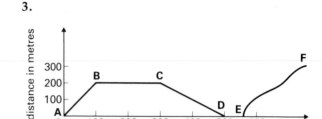

time in seconds

The above diagram represents the motion of somebody walking along a straight road.
  i Between which two points on the graph is the walker travelling with uniform speed away from his starting point? Calculate the value of the uniform speed at this stage.
  ii Between which two points is the walker travelling with uniform speed towards his starting point? Calculate the speed at this stage.
  iii Between which points is the walker stationary? For how long a time over the whole motion shown is the walker stationary?
  iv Between which points is the walker travelling with non-uniform speed?
  v How far does the walker go in the first 50 seconds?
  vi How far does he go in the second 50 seconds?
  vii If he had continued at this rate, how long would he have taken to get 1000 metres from his starting point?   (M.R.E.B.)

4. a What do you understand by the term "acceleration"?
  b The diagram shows a velocity-time graph for a car on a test run:

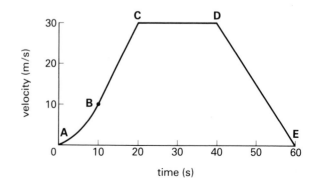

time (s)

Describe the motion of the car during the periods represented by
i AB   ii BC   iii CD   iv DE
and give *one* reason for your answer to each of these.

c At what time does the car have a speed of 25 m/s?

d i What is the acceleration during the part **BC?**
ii What is the acceleration during the part **DE?**
iii Why is this acceleration negative?
(A.L.S.E.B.)

5.
i An object with no unbalanced forces acting on it can be in one of two possible states of motion. What are these two states?
ii If an unbalanced force does act on the object, what effect does it produce?
iii A man stands at the stern of a rowing boat, throwing bricks backwards out of the boat. Explain why the boat moves forwards.
iv Suggest TWO ways by which he could move the boat faster by this process.   (W.M.E.B.)

6. a The following are two statements about how we get electrical energy from coal:—
i Coal is burned to make steam to drive turbines.
ii Turbines turn generators.
Outline the energy changes which take place in statement **i** and statement **ii**.
b Describe how the energy was stored in the coal originally.   (E.M.R.E.B.)

7. A boy of weight 500 N climbs 10 metres from water level up to a diving board.
i What is his gain in potential energy? State the units.
ii What is the source of the extra potential energy?
iii The boy steps off the diving board. What will be his kinetic energy just before he hits the water?
What fundamental scientific principle does this illustrate?   (W.M.E.B.)

8. A boy on a bicycle started from rest at a point A on a hill and rode down without pedalling or using his brakes. Point A is 10 m above the bottom of the hill, measured vertically as shown. The boy and his bicycle together weigh 360 N.

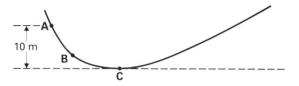

i What kind of energy have the boy and his bicycle got at A?
ii What kind of energy have the boy and his bicycle got at B?
iii What kind of energy have the boy and his bicycle got at C?
iv Copy the diagram and mark on it the point D to which he will rise on the other hill. (Ignore the effects of friction).
v Calculate the energy of the boy and his bicycle at point A.
vi If the boy started from rest at C and rode up to A in 20 s, calculate his power.   (E.M.R.E.B.)

9. a Draw a pulley system with a velocity ratio of three.
b i In a block and tackle system of velocity ratio 5 an effort of 500 N is required to raise a certain load. If the system is 60% efficient find the value of the load.
ii Give one possible reason for the efficiency being less than 100%.   (N.W.R.E.B.)

10. Explain the terms "mechanical advantage", "velocity ratio", and "efficiency" for a machine. Calculate their values for a machine given that when the effort of 30 N moves 40 cm, the load of 100 N moves 10 cm.   (M.R.E.B.)

11. The diagram shows a brick of weight 20 N being pulled up a flat plank by a steady force of 10 N.

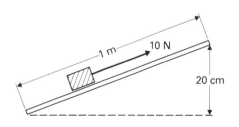

i What is the velocity ratio of the system?
ii What is its mechanical advantage?
iii What is its efficiency?   (S.R.E.B.)

37

Pressure: spreading the force

There is a large force on the unfortunate stunt cyclist's assistant, shown in the photograph below. But because he has covered himself with a board, the large force is spread out over a large area – this stops him from being crushed:

The effect of the force varies with the area on which it presses. The *pressure* that the force exerts varies with the area. For a particular force, the pressure is greater when the area is smaller.

Pressures are *compared* by finding the force that acts on one square metre of area.

Pressure is defined as *force per unit area*. Forces are measured in newtons (N) and areas are measured in square metres (m²), so pressures are measured in newtons per square metre (N/m²).

$$\text{pressure} = \frac{\text{force}}{\text{area}} \text{ newtons per square metre}$$

$$P = \frac{F}{A} \quad \text{N/m}^2$$

Pressure in liquids

The situation shown in the photograph is only one example of the fact that all objects exert a pressure on the surface that supports them. In the same way, a cylinder filled with water exerts pressure *downwards* on the bottom of the cylinder. But because liquids can flow, the water needs support on the sides as well as underneath it to hold it in place – it exerts a pressure *sideways* on to the walls of the cylinder, as shown in the diagram:

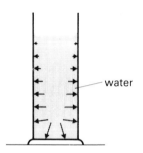

water

Each 'layer' of liquid is also supporting the layers above it. It supports them by pressing up on them – pressure also acts *upwards*.

The pressure on the sides is not the same all the way down. Each depth of water is supporting all the water above it. So all depths, except for the very surface, are under pressure. The greater the depth, the greater the pressure.

The pressure increases *only* with the depth. In other words, the pressure at any one depth is the same, no matter what the shape of the container. To summarise:

1. **Pressure at any one depth is the same.**
2. **Pressure increases with depth.**
3. **Pressure acts in all directions.**

Demonstrating pressure in liquids

All these effects may be shown in laboratory experiments.

Pressure at any one depth is the same. If holes are drilled all the way round a can at the same level, and the can is filled with water, the water shoots out from each of the holes to the same extent as shown in the diagram at the top of the next column:

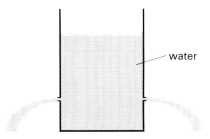

Pressure increases with depth. If holes are drilled at various heights on a tall can, and the can is filled with water, water shoots out furthest from the lowest hole and least from the top one:

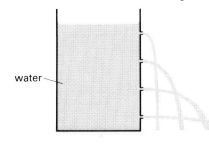

Pressure acts in all directions. A syringe with a bulb-shaped end as shown below, will demonstrate this. The bulb is pierced with small holes all over:

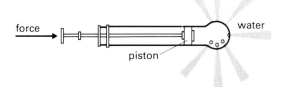

When it is filled with water and the plunger is forced in, the water shoots out equally in all directions.

Problems from increasing pressure with depth. The fact that the pressure in a liquid increases with its depth has to be taken into account in various ways, of which two examples are given here.

The construction of dams. The engineers who built the dam shown at the top of the next column had to take into account the increase of pressure with depth. The dam wall is built thicker at the bottom than the top, to withstand the greater pressure.

Deep sea diving. The diver shown in the photograph on page 41 is breathing compressed air. As he goes deeper the cylinders have to provide air to his lungs at a greater pressure – otherwise, his chest would be crushed by the water pressure. At these high pressures nitrogen from the air gradually dissolves in the blood. If the diver comes back to the surface too quickly, the nitrogen forms bubbles in the bloodstream, and causes "the bends", a condition similar to a heart attack. The diver has to surface very slowly, in order to prevent this.

Exercises
1. How are pressures calculated?
2. What is the definition of pressure and what are its units?
3. Complete the following sentence, using the words below.

> The pressure exerted by a solid only presses *, but that exerted by a liquid acts *. A force will produce a higher pressure when acting on a * surface, than when it acts on a * one.

(small, in all directions, downwards, larger)
4. Describe three experiments that show the way in which the pressure in liquids acts.
5. Explain how one of the problems caused by increasing pressure with depth is overcome.

More about pressure in liquids

The pressure due to a liquid increases with its depth. This is used to measure pressure, to calculate pressures, and to compare the densities of two liquids.

Measuring pressure: the manometer

The manometer consists of a U-tube which contains a liquid such as water. The gas supply pressure may be measured by connecting one side of the U-tube to the gas tap. The pressure of the gas forces the liquid round the tube, so that the liquid in one side is a height h above the liquid in the other. The gas pressure is then equal to the pressure due to the depth h of liquid:

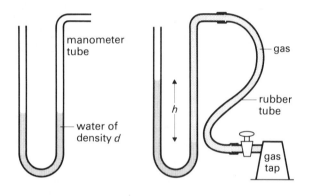

If manometers containing liquids of different densities are connected to the same gas supply it is found that the values of h are different. The liquid of higher density is pushed up a much smaller distance by the same pressure. Mercury, which has a relative density of 13.6 is used for measuring high pressures: it will only rise one 13.6th as high as water under the same pressure.

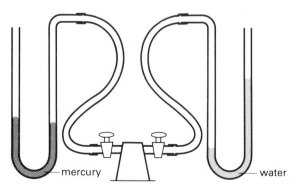

Calculating pressure: formula for liquid pressure

The pressure in a liquid depends on:

1. **The depth:** pressure increases with depth.
2. **The density of the liquid:** pressure at a particular depth is higher in a more dense liquid. This is because there is a greater weight of the more dense liquid pressing down.

The pressure also depends on the acceleration due to gravity. A strong gravity field will increase the weight of the liquid, and therefore its pressure.

These effects are combined in one formula to work out the pressure at a particular depth in a liquid.

$$\text{pressure} = \text{depth} \times \text{density} \times \frac{\text{gravity}}{\text{acceleration}}$$
$$P = h\,d\,g \quad \text{N/m}^2$$

Comparing densities of liquids

If columns of two different liquids are supported by the same pressure, then the height of each column indicates the density of its liquid. The taller column of liquid has the lower density. This situation can be set up using two different pieces of apparatus.

1. **Hare's apparatus.** This is shown in the diagram below. Some of the air above the two liquids is removed by sucking it out through the tap. The atmospheric pressure now pushes on both columns of liquid until it is balanced by the pressure due to the column of liquid, plus the pressure above the liquid, as shown in the next diagram:

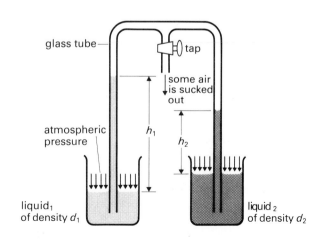

The liquid of lower density is forced to a greater height up the tube. The densities of the liquids are compared by comparing the heights of the liquid columns:

$$\frac{\text{density of liquid}_1}{\text{density of liquid}_2} = \frac{\text{height of liquid}_2}{\text{height of liquid}_1}$$

$$\frac{d_1}{d_2} = \frac{h_2}{h_1}$$

If the density of one liquid is known then the other density can be found. The height of the column above the liquid surfaces in each beaker is measured. The figures are then substituted in the formula.

2. U-tube. This method can only be used for liquids that do not mix. The two liquids are placed in the U-tube. A taller column of the less dense liquid is needed to balance a shorter column of the more dense liquid:

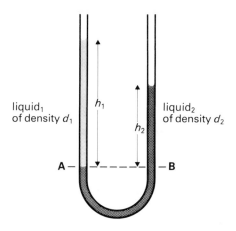

liquid$_1$ of density d_1 h_1 liquid$_2$ of density d_2 h_2

A — — — — — B

The liquid in the tube below the line **AB** is compressed equally by both columns. If the pressure due to each liquid column were not the same, the relative levels would adjust until they were. Therefore, the pressure at **A** is the same as the pressure at **B**, and the heights of the liquid columns must be measured from the level **AB**. The densities of the two liquids may then be compared by using the same formula:

$$\frac{\text{density of liquid}_1}{\text{density of liquid}_2} = \frac{\text{height of liquid}_1}{\text{height of liquid}_2}$$

$$\frac{d_1}{d_2} = \frac{h_2}{h_1}$$

Exercises

1. What is the pressure at a depth of 8 m in a liquid of density 800 kg/m³? Take g as 10 m/s².

2. Explain, with the aid of a diagram of the apparatus used, how you would measure the gas pressure in the laboratory.

3. In the Hare's apparatus diagram, $h_1 = 0.5$ m and $h_2 = 0.4$ m. Work out the density of liquid **1** assuming that liquid **2** has a density of 1000 kg/m³.

4. In the U-tube diagram $h_1 = 0.3$ m and $h_2 = 0.2$ m. Work out the density of liquid **1** assuming that liquid **2** has a density of 900 kg/m³.

5. The diver shown in the photograph is breathing compressed air, at a pressure of 200 000 N/m². The water density is 1 000 N/m² – what is his depth? Take g as 10 m/s².

Atmospheric pressure

The Earth is flooded with a deep 'sea' of air. Above that 'sea' is empty space. Any empty space is called a *vacuum*.

It is the Earth's gravitational force which keeps the 'sea' of air around the Earth. At the bottom of the sea of air, on the Earth's surface, there is a pressure due to the weight of all the air pressing down. This pressure is called the *atmospheric pressure* and it has a value of about 100 000 N/m².

This is a fairly strong pressure, but our bodies are designed to cope with it. Air is taken into our bodies so that the pressure inside is the same as the pressure outside:

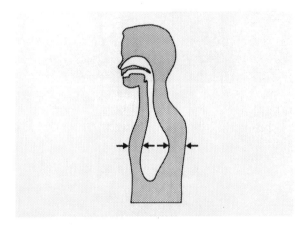

There is a balance between the force pressing in and the force pressing out. What happens when these two forces do not balance can be shown by pumping the air out of an oil drum. As the air is removed from inside the drum, the air pressure crushes it:

Magdeburg hemispheres

Another method of demonstrating the strength of the atmospheric pressure uses *Magdeburg hemispheres*. These are two hemispheres which fit tightly together so that the air inside them can be pumped out. They are made specially strong so that they do not collapse like the oil drum:

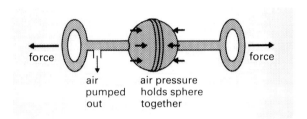

When the air is pumped out from inside the spheres, it is found that they cannot be pulled apart. The atmospheric pressure holds them together.

Measuring atmospheric pressure

The atmospheric pressure can be measured by pumping the air out of a long vertical tube, whose bottom end is in a dish of mercury. As the pump gradually removes the air, the atmospheric pressure pushes the mercury up the tube. Eventually a point is reached when all the air has been removed from the tube and the atmospheric pressure can push the mercury up no further. The pressure due to the weight of the column of mercury is then equal to the atmospheric pressure. It is found that the atmospheric pressure will support a column of mercury about 0.76 m high:

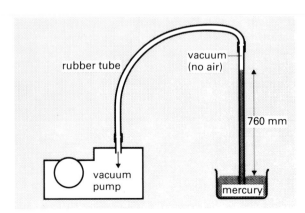

Calculating the atmospheric pressure. The atmospheric pressure is exactly equalled by the pressure due to the weight of the column of mercury. The pressure due to the column of mercury, and therefore the atmospheric pressure, can be calculated from the formula:

$$\text{pressure} = \frac{\text{height of}}{\text{column}} \times \frac{\text{density of}}{\text{mercury}} \times \frac{\text{gravity}}{\text{acceleration}}$$

$$= h \, d \, g \quad \text{N/m}^2$$

Typically:

$$\text{pressure} = 0.76 \times 13\,600 \times 10 \text{ N/m}^2$$

$$= 103\,360 \text{ N/m}^2$$

The atmospheric pressure is therefore just over $100\,000 \text{ N/m}^2$ at the surface of the earth.

The mercury barometer

The simple mercury barometer is made from a one-metre long tube that is sealed at one end. It is filled with mercury, and the open end is inverted into a mercury-filled dish. The level of the mercury in the tube drops until the height of the column is 0.76 m. A vacuum is formed in the part of the tube which is above the mercury:

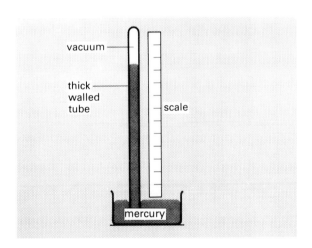

vacuum

thick walled tube

scale

mercury

For simplicity atmospheric pressure is often measured in millimetres of mercury. The actual height of the column varies with the weather. The air pressure is usually low when the weather is rainy – the mercury level falls. The pressure is high when the weather is clear – the mercury level rises again.

Tilting the barometer. When the tube is tilted, the mercury moves up the tube until the vertical height of the top of the mercury column is again 0.76 m above the mercury in the dish. If the tube is tilted so that the top is below a height of 0.76 m the mercury will fill the tube:

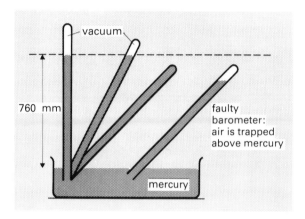

vacuum

760 mm

faulty barometer: air is trapped above mercury

mercury

If some air has accidentally been trapped above the mercury, its pressure will prevent the mercury rising to the top of the tube. Tilting the tube is therefore used to check if the barometer is faulty.

Exercises

1. Explain what happens to a sealed can when the air is removed from inside it.

2. Why do Magdeburg hemispheres not pull apart? What would you do in order to enable them to pull apart?

3. What happens to a column of mercury when the air pressure above it is reduced?

4. A sealed 2 m tube full of mercury is inverted in a dish of mercury. About how long is the vacuum space above the mercury?

5. What is the atmospheric pressure in N/m² when it will support a column of mercury 0.8 m high? (take $g = 10$ m/s² and density of mercury as 13 600 kg/m³).

6. Explain how you would make a simple mercury barometer, and how you would test whether the barometer was faulty. Draw diagrams to illustrate your answer.

Measuring pressure

In the vacuum of space above the Earth the air pressure is zero. The air pressure gradually increases as the surface of the Earth is approached, and is at maximum at the Earth surface. This means that a special barometer called an *altimeter* can have a scale of metres above sea level marked on it:

The aneroid barometer

The air pressure does not remain the same at ground level, but changes from day to day. It is often the case that when the pressure is higher than the normal, the skies are clear and the weather is fine. On the other hand, when the pressure is low it is often cloudy and the weather is rainy. In this way, changes in the air pressure can be used to predict changes in the weather. The most commonly used type of barometer to do this is *the aneroid barometer:*

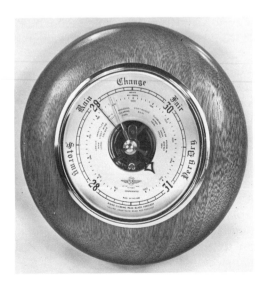

How the aneroid barometer works

The main part of the aneroid barometer is a sealed, circular metal box with ridges round it. It is partially evacuated – some of the air inside it has been pumped out – but a strong spring stops it from squashing too much. The box will respond to changes in the air pressure by compressing or expanding very slightly:

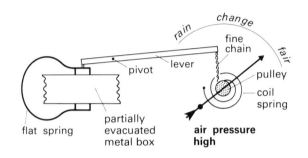

When the air pressure increases, the force on the box increases causing it to compress slightly. This pulls on the fine chain, and turns the pointer clockwise.

When the air pressure decreases the box expands. The fine chain goes slack, so the coil spring turns the pointer back anticlockwise:

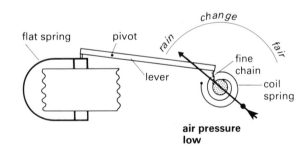

The bourdon gauge

The aneroid barometer is useful for detecting very small changes in atmospheric pressure, but when much larger pressure changes are being measured a *bourdon gauge* is used. The oil pressure gauge in a car, and the gauges on top of the gas cylinders in the chemistry laboratory are bourdon gauges:

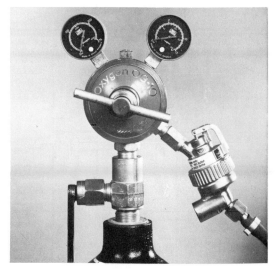

When the pressure of the liquid or gas in the tube increases, the tube straightens out slightly. The end of the tube goes up and moves the gear segment round which in turn moves the pointer:

The bourdon gauge works on the same principle as the party toy shown at the top of the next column. Blowing through it increases the pressure in the paper tube, which makes it unroll.

The working bourdon gauge consits of a curved metal tube which is connected to a pointer by two gears as shown below:

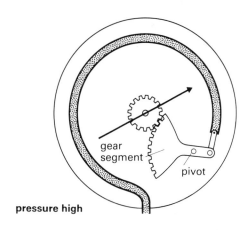

gear segment

pivot

pressure high

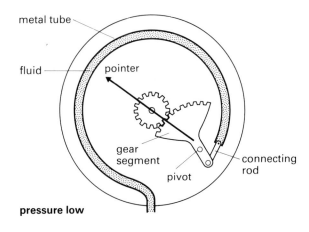

metal tube

fluid

pointer

gear segment

pivot

connecting rod

pressure low

Exercises
1. How can a device that measures pressure measure height?
2. Explain with diagrams how the weather-forecasting aneroid barometer works.
3. Why can aneroid barometers not be used to measure large pressure changes?
4. Explain with diagrams how the bourdon gauge works.

Pumps

Pumps are machines which force liquids or gases from one place to another. The simplest type of pump is the hypodermic syringe, used by doctors when giving injections. It consists of a cylinder with a closely fitting piston; the end of the cylinder has a hollow, very fine needle mounted into it.

The syringe is filled by pushing the cylinder down, forcing the air out of the cylinder as shown in diagram **1**:

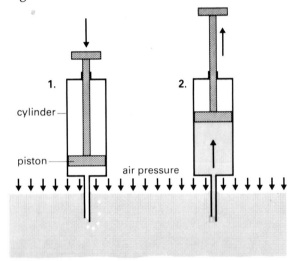

The piston is then drawn upwards, as shown in diagram **2**. A vacuum is created underneath the piston so the air pressure forces the liquid up the hole in the centre of the needle, into the cylinder. The liquid can now be squirted out by pushing the piston down.

The syringe is a simple pump because it does not need any *valves*. A valve will allow a liquid or gas to flow in one direction, but not the other. The rest of the pumps in this section do need valves.

Valves are usually made from a large ball bearing, held into a metal cone by a spring, or by gravity. When the fluid flows in one direction, it pushes the ball out of the cone, and can therefore continue to flow. When it attempts to flow the other way, it pushes the ball into the cone, causing the ball to stop the flow.

The lift pump
This type of pump is used in caravans to pump water from the storage tank to the sink, and in small boats to pump out any water which has leaked into them.

Operation of the pump. The pump consists of a piston and a cylinder, and two valves, one of which is mounted in the piston, the other at the base of the cylinder.

On an *upstroke* – when the piston is lifted – two things happen. First, any water already above the piston is lifted out through the pipe. The weight of the water keeps the piston-valve shut. Second, a vacuum forms below the piston. The atmospheric pressure forces water through the bottom valve to fill this space:

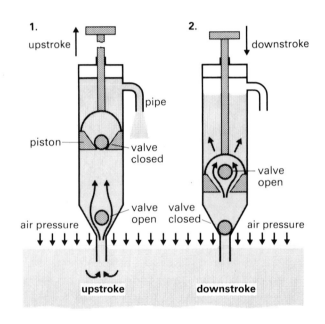

On the *downstroke* – when the piston is lowered – the weight of the water closes the bottom valve. The water flows through the piston-valve as the piston moves down through it. The water is now above the piston, where it is ready to be lifted out on the next upstroke.

As the piston is lifted, it is the air pressure that forces the water into the cylinder underneath it. Atmospheric pressure can only support a column of water 10 m high. This means that the pump cannot be more than 10 m above the surface of the water that it is being used to pump.

The force pump

This type of pump is used to develop very high pressures.

Operation of the pump. On the upstroke, a vacuum forms under the piston, and atmospheric pressure forces water through the valve at the bottom of the cylinder, to fill this space. Water from the side cylinder is prevented from coming into the space by the side valve. In this way, the cylinder is filled:

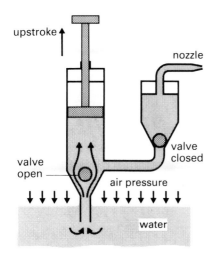

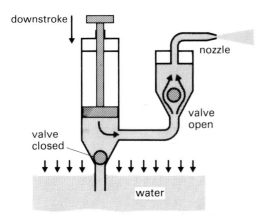

On the downstroke of the piston, the valve at the bottom of the cylinder closes. The water is forced out through the side valve with a pressure that depends only on the amount of force applied to the piston.

The bicycle pump

This is a force pump that pumps air. One of the valves necessary for its operation is inside the bicycle tyre, and the second is built into the design of the oiled leather cup washer:

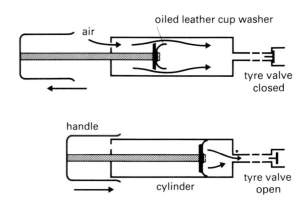

As the handle of the pump is pulled out, air slips past the washer into the body of the pump. As the handle of the pump is pushed in, the air catches on the cup washer and forces the leather against the cylinder making an airtight seal. As the handle is pushed in further, it forces the air through the valve mounted in the rim of the wheel, and into the tyre. Once in the tyre, the air cannot escape through the valve.

Exercises

1. What is the simplest type of pump? Explain with diagrams how it works.
2. Explain with diagrams how the lift pump works.
3. Why can the lift pump raise water by about a maximum of 10 m?
4. Explain with diagrams how the force pump works.
5. Explain with diagrams how the bicycle pump works. What will happen if the leather washer is allowed to dry out?

Hydraulics

Many machines such as the rams that lift tipper-trucks, are operated by *hydraulic pressure*. Hydraulic systems work by connecting cylinders of different diameters with a tube, which transfers the pressure between them. This section explains how small forces can be converted into large ones using hydraulic systems.

Using the apparatus shown below it is found that a small force is able to balance a large one.

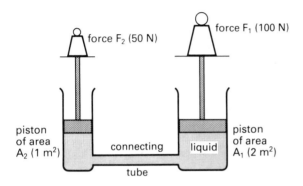

The large force on the large piston makes a pressure under it:

$$\frac{\text{pressure under}}{\text{large piston}} = \frac{\text{force}}{\text{area}} = \frac{100}{2} = 50 \text{ N/m}^2$$

On the *small* piston a *small* force is needed to make *the same* pressure:

$$\frac{\text{pressure under}}{\text{small piston}} = \frac{\text{force}}{\text{area}} = \frac{50}{1} = 50 \text{ N/m}^2$$

Putting this mathematically:

$$\frac{\text{large force}}{\text{large area}} = \frac{\text{small force}}{\text{small area}}$$

$$\frac{F_1}{A_1} = \frac{F_s}{A_s}$$

If the small force is increased slightly it will overcome the large force and raise it.

If it does so, the small force has to move a long way to make the large force move a short distance. To move the large force a *large* distance, the small cylinder is made into a pump. The large piston can then be moved a long way by many strokes of the small piston.

The hydraulic jack

The hydraulic car jack uses this principle to create large forces.

In the simple jack, the lever operates the small piston of a force pump. Each stroke pumps another charge of oil into the space under the large piston, thus forcing it up a small amount each time:

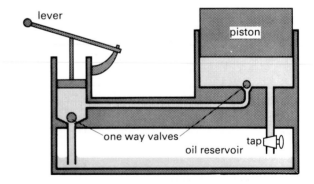

The large piston is lowered by opening the tap which allows the oil under it to drain back into the reservoir.

Car brakes

The braking system on all modern cars is hydraulic. Force from the driver's foot is applied to the brake shoes on each wheel:

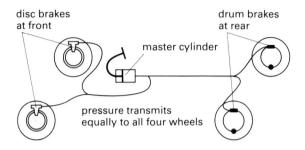

When the brake pedal is pressed, pressure is built up in a *master-cylinder*. The pressure is transmitted uniformly through a system of pipes to *slave cylinders,* so that the braking force is evenly applied. There are two types of brakes on modern cars:

Drum brakes. In this type, a brake shoe is forced against a brake drum which is attached to the wheel. Friction between the two forces the wheel, and thus the car, to slow down.

When the brake pedal is pushed down, fluid from the master-cylinder is pushed along the pipes to each wheel. In the slave cylinder, the pistons are forced apart, pressing the brake shoes against the drum:

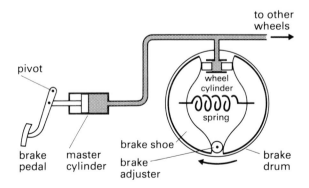

When the brake pedal is released, the spring pulls the brake shoes off the drum, and the pistons in the slave cylinders are pushed back together. This forces the fluid back along the pipes to the master-cylinder.

Disc brakes. These work in a similar way. Hydraulic pressure from the master-cylinder forces two *brake pads* against a steel disc which is fixed to the wheel:

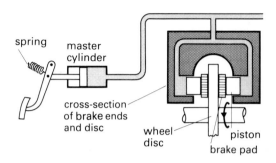

Friction between the brake pads and the disc slows the wheel down, and thus slows the car also. When the brake pedal is released, the piston in the master-cylinder is drawn back releasing the pressure. The brake pads no longer press against the disc.

Exercises

1. Complete the following sentence, using the words given below.

In a hydraulic system, a * force travelling a large distance can be changed into a * force moving a small distance. This is done by moving a * in a small area cylinder so that it forces * into a tube of larger area.

(small, liquid, large, piston).

2. In a car jack, the area of the small piston is 10 cm² and the area of the large one is 40 cm². If the maximum force that can be applied to the jack is 500 N, what is the maximum load it can raise?

3. Explain with the aid of a diagram, how the drum brake works.

4. Explain with the aid of a diagram, how the disc brake works.

Questions on chapter 3

1. i Explain why it is necessary to wear snow shoes to prevent one from sinking in the snow.
ii Calculate the pressure exerted on snow by a man of weight 800 N when he wears snow shoes having a total area of 0·5 m² in contact with the snow. State clearly the units you are using.
(W.J.E.C.)

2.

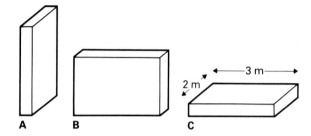

i The same concrete block is shown resting on the ground in three positions. In which position is it exerting the greatest pressure on the ground?
ii Work out the area of the largest face from the dimensions on the diagram.
iii What pressure does the block exert on the ground in diagram **C** if the weight of the block is 36 000 N? (S.W.E.B.)

3. Copy the following diagram into your book.

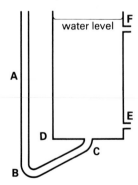

i On the diagram that you have drawn, mark the level of the water in tube A.
ii Draw the jets of water from spouts E and F
iii State which letter on the diagram is nearest the point of highest pressure. (Y.R.E.B.)

4.

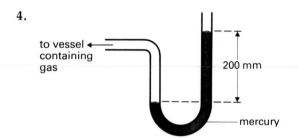

A simple pressure gauge is connected to a vessel containing gas. The gauge contains mercury. The atmospheric pressure is 760 mm of mercury.
i What is this type of pressure gauge called?
ii What is the pressure of the gas supply?
iii What liquid would you use instead of mercury for small pressure differences?
iv Explain your answer to **iii**. (E.M.R.E.B.)

5. A glass tube, open at both ends, is placed in water as shown in diagram **1**. The top is then sealed and the tube lifted as shown in diagram **2**. Finally, the tube, still sealed, is lowered as shown in diagram **3**.

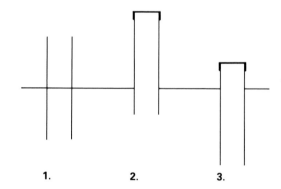

Copy the diagrams into your book.
i Show on your copies of diagrams **2** and **3** the approximate positions of the water levels.
ii Inside which tube is the pressure of the air least? (E.A.E.B.)

6. i Describe with the aid of a labelled diagram, how an aneroid barometer measures atmospheric pressure.
ii Draw a diagram and carefully explain how the atmosphere holds a rubber sucker on a smooth surface. Explain what happens if the surface is roughened. (E.A.E.B.)

7.

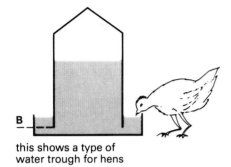

this shows a type of
water trough for hens

i Explain how the water is held up inside the water trough.
ii Describe what will happen when the hen has drunk enough water to bring the water level in the outside of the trough below level **B.**   (Y.R.E.B.)

8.

i Name the instrument shown in the diagram.
ii What is it used to measure?
iii What is the liquid it contains?
iv What occupies the space labelled **s**?
v What is the approximate value, in centimetres, of the height *h*?   (N.W.R.E.B.)

9. i Define the term "pressure" and give the unit in which it is measured.
ii The wall of a large dam tends to get progressively thinner towards the top. Explain the reasons for this design.
iii A motor vehicle weighing 10 000 N is supported on the road on four tyres. If the area of each tyre in contact with the road surface is $1/100$ m² ($0 \cdot 01$m²), calculate the pressure in each tyre.   (N.W.R.E.B.)

10.

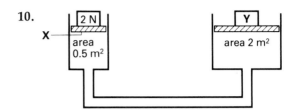

The diagram shows a hydraulic press.
i What is the pressure at **X**?
ii What is the load at **Y**?
iii State one other application of the hydraulic principle.   (W.M.E.B.)

11. The diagram represents part of the hydraulic brake system of a car.

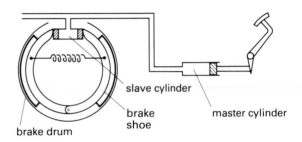

i With the aid of the diagram, explain how such a system operates. Be careful to explain fully the part played by fluid pressure in the system.
ii Explain the effect on braking of an air bubble in the hydraulic brake fluid.   (W.M.E.B.)

12. Draw a fully labelled diagram of a hydraulic jack and explain its action. Explain why oil is used as the fluid in a hydraulic machine.   (W.J.E.C.)

Kinetic theory — the three states of matter

The kinetic theory gives scientists a way of explaining why the three states of matter (solids, liquids and gases) behave the way they do.

The kinetic theory of matter
The kinetic theory is based on the following statements:

1. All matter is made up of small particles called molecules.

2. The molecules are in constant rapid motion or vibration.

3. The higher the temperature, the faster the molecules are moving.

One of the first things that can be done with the kinetic theory is to explain the difference between *heat* and *temperature:*

Heat. This is a form of energy, measured in joules (J). When heat energy is supplied to an object, its molecules vibrate or move faster.

Temperature. This indicates the speed of the molecules. Temperature is often measured in degrees Celcius (°C). As heat is supplied to an object, its temperature goes up.

Solid state
The molecules of a solid are arranged in a regular pattern called a lattice:

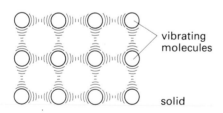

The molecules are vibrating backwards and forwards but are held in their positions by the lattice forces and cannot move around. For this reason solids have a definite shape.

Melting. As the temperature rises to the melting point, the vibrations become so violent that the order of the lattice breaks down. The molecules become free to move around – the solid becomes a liquid. The temperature remains steady whilst the

A common substance in solid, liquid, and gas state.

solid melts – the heat supplied is being used to free the molecules from forces in the lattice. The heat used to change the solid to a liquid is called the *latent heat of fusion.* When the liquid solidifies, the same latent heat is given out.

Liquid state
In a liquid the molecules are about the same distance apart as in a solid but are free to move around. This means that a liquid will pour, and adopt the shape of the container into which it is poured. There is still a force of attraction between the molecules making them stay together, so a liquid has a definite surface:

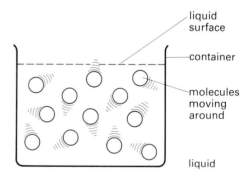

Boiling. When the temperature of a heated liquid reaches the boiling point, water molecules leave the liquid and become a gas. The temperature remains steady as the liquid boils. This is because the heat being supplied is used to free the molecules from the forces of attraction that hold them in the liquid. The necessary heat is called *latent heat of vaporization.* When the gas condenses into a liquid, the same latent heat is given out.

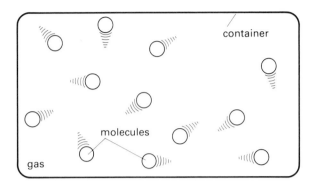

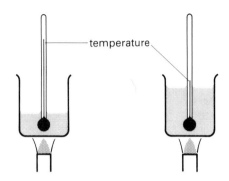

21

Gaseous state

In a gas the molecules are much further apart than in a liquid. They move a lot faster, as shown in the diagram above.

There is almost no force of attraction between the molecules. They have a lot of energy of movement, so they move out to fill the container into which the gas is placed. The pressure of a gas is caused by the force of the molecules hitting the walls of the container.

Absolute zero of temperature

When a group of molecules go from being solid to liquid to gas, they are being given heat – their speed and temperature are increasing. In the same way, as heat is removed from a group of molecules, their speed and temperature drops. The kinetic theory predicts that when so much heat has been removed from them that they do not move at all, a "lowest possible temperature" is reached. This temperature is called *absolute zero* and is −273 °C. Nothing can get colder than this temperature.

Scientists frequently measure temperature in degrees Kelvin (K). This scale of temperature is called the *absolute scale of temperature* as it starts at absolute zero. The table below gives both the Kelvin and Celsius temperature for important reference points:

absolute zero	0 K	−273 °C
freezing point of water	273 K	0 °C
boiling point of water	373 K	100 °C

The Kelvin and Celsius temperatures are related by the formula:

degrees K = degrees C + 273

An experiment on heat and temperature

Two beakers that are heated with the same bunsen for the same length of time, receive the same amount of heat, as shown above.

The one with less water in heats up faster. This happens because the beaker containing more water receives the same amount of heat, but this has to be shared amongst more molecules. The increase in speed of each molecule is therefore less – the temperature increase is less too.

Exercises

1. What are the three assumptions of the kinetic theory?
2. Draw a diagram showing the molecules in a solid. How does the kinetic theory explain the fact that a solid has a definite shape?
3. Draw a diagram showing the molecules in a liquid. How does the kinetic theory explain that a liquid will pour and that a liquid has a definite surface?
4. Draw a diagram showing the molecules in a gas. How does the kinetic theory explain gas pressure and the fact that a gas fills the container into which it is placed?
5. Copy out the following sentences filling in the missing words from the lists.

The * of a substance remains * whilst it is * state. (constant, changing, temperature)

The * needed to * or * a substance is called the * heat. (latent, heat, vaporize, melt)

Kinetic theory — evaporation, Brownian motion, and diffusion

The kinetic theory must be able to explain all the various properties of solids, liquids and gases. This section shows how it explains three different effects.

Evaporation

A drop of methylated spirit, even when placed on a hard surface, slowly vanishes. The liquid methylated spirit has become a gas – it is said to have *evaporated*. Evaporation differs from boiling in two ways:

1. Boiling occurs at one particular temperature.

2. Boiling takes place throughout the volume of the liquid. Bubbles of gas are formed in all parts of it. Evaporation only takes place at the surface and no bubbles are formed:

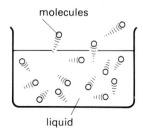

boiling   evaporation

Kinetic theory of evaporation

To explain evaporation, the kinetic theory suggests that only the *average* speed of the molecules in a liquid at a particular temperature remains steady. The molecules are continually colliding with one another, so that some end up with a speed higher than the average, and some lower. Fast moving ones near the surface may be going fast enough to break free from the surface and become a gas:

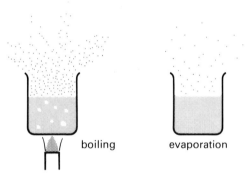

molecules

liquid

Only molecules which are near the suface and are moving with higher than average speed can evaporate. As the liquid loses its fastest molecules by evaporation, the average speed of the molecules becomes slower. In other words, the temperature of the liquid should become lower as evaporation occurs. This can be tested by a simple experiment:

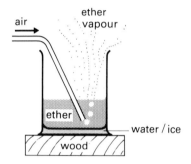

air   ether vapour

ether

water / ice

wood

The ether in the beaker is made to evaporate quickly by blowing air through the tube. The water under the beaker becomes so cold that it freezes, sticking the beaker to the wood. Evaporation does cause cooling; this is consistent with the theory.

Factors increasing evaporation

Three main factors are involved:

Temperature. When the temperature is higher, the average speed of the molecules is higher. Therefore, there will be more molecules moving fast enough to escape.

Surface area. A large surface area means that there are more molecules near the surface – more are able to escape.

Wind. The wind drives the evaporated molecules away. None can return to the liquid.

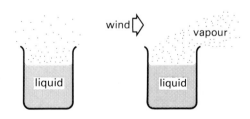

wind   vapour

liquid   liquid

22

Brownian motion

Dust particles floating in the air are made visible by a shaft of sunlight coming through partly closed curtains. The dust particles appear to be jogging about in all directions. This effect is called Brownian motion, and it can be explained by the kinetic theory.

The air molecules, although too small to be seen, are in continual rapid motion. The air molecules keep hitting the dust particles, knocking them in all directions. The dust particles move much slower than the air molecules because they are much heavier.

Diffusion

When one substance gradually spreads through another, it is *diffusing* through it. Diffusion can be explained using the kinetic theory:

In liquids. A purple crystal of potassium permanganate is placed at the bottom of a beaker of water. As it dissolves, the purple colour gradually spreads throughout the water:

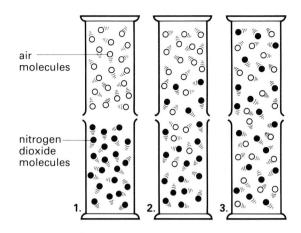

Both the nitrogen dioxide molecules and the air molecules are in constant rapid motion. Some air molecules gradually move down into the space occupied by the nitrogen dioxide, and some nitrogen dioxide molecules move upwards, as shown in diagram **2**. This process continues until both types of molecules are evenly spread throughout as shown in diagram **3**.

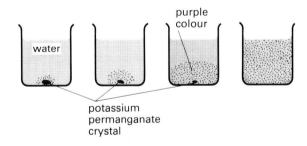

The kinetic theory explanation is that both the water molecules and the dissolved potassium permanganate molecules are in constant motion. Therefore the potassium permanganate molecules will eventually spread throughout the water.

In gases. To illustrate this, a gas jar is filled with the heavy brown gas nitrogen dioxide, and another gas jar filled with air is placed on top of it as shown in diagram **1** at the top of the next column.

The brown colour of the nitrogen dioxide gradually spreads upwards until it is evenly spread throughout the two gas jars:

Exercises

1. Explain what is meant by *evaporation* and state how it differs from boiling.
2. Describe an experiment to show that evaporation causes cooling. How does the kinetic theory explain this?
3. What three factors increase evaporation?
4. Why do clothes on a washing line dry faster on a warm breezy day than on a cold still day?
5. Why do dust particles in still air appear to jog about? What is this effect called?
6. Explain what happens when a blue crystal of copper sulphate is placed in a beaker of warm water. What is this effect called?

Surface properties of liquids

The force of attraction between molecules causes the surface of a liquid to behave like an elastic skin; it also draws liquids up fine holes such as those in blotting paper.

Surface tension

The photograph shows a drop of water forming on a tap. The diagram next to it illustrates the kinetic theory explanation of why it is forming into a circular shape:

 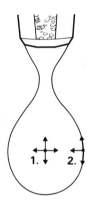

There is a force of attraction between all the liquid molecules. Molecule **1**, which is surrounded by other molecules, is pulled equally towards all of them. Molecule **2** at the surface has molecules pulling sideways and inwards, but not outwards. Overall, all the molecules at the surface are pulled inwards. The surface behaves as if it has an elastic skin pulling it in to the smallest volume for its size. This shape is that of a sphere. The effect of the "skin" is called *surface tension*.

Small droplets of water resting on a surface such as paper often form spheres. If a small quantity of detergent is added to the water, then the droplets wet the paper and soak in:

The droplets soak in because the detergent has reduced the surface tension of the water. Any impurity reduces surface tension. Raising the temperature of a liquid also has the same effect. Clothes are washed in hot water containing detergent to combine both effects. This makes sure that the water thoroughly wets the fibres of the cloth and removes the dirt.

Soap films demonstrate the elastic behaviour of a liquid surface. When a soap film is stretched across a wire loop, a piece of cotton tied to the loop will float in the film as shown in diagram **1**. If the film is broken at one side the surface tension in the other side pulls the cotton taut as in **2**:

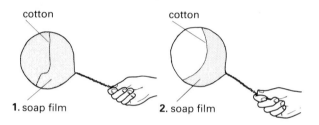

Some creatures have adapted to use the surface tension skin of water. The pond skater walks on the surface "skin" of water, and mosquito larvae hang from it:

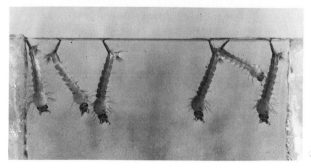

23

Capillary action

When a narrow tube is placed in a liquid, the level of the liquid inside the tube is often higher or lower than that outside the tube. This effect is called *capillary action,* and it can be explained using the kinetic theory. The forces between molecules are divided into two types:

Cohesion. The attraction force between molecules of the same substance is a *cohesion* force.

Adhesion. The attraction force that exists between molecules of different substances is called an *adhesion* force.

The way in which liquids act, when they come into contact with solids, depends on which of these two forces is stronger.

When a narrow glass tube is placed in a dish of water, the level of the water inside the tube rises above the level outside:

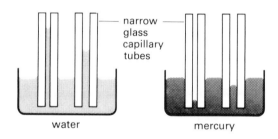

narrow glass capillary tubes

water   mercury

The *adhesion* force between water molecules and glass molecules is greater than the cohesion force between water molecules. The water therefore rises up the tube so that more water molecules can be in contact with glass molecules.

If the water is replaced by mercury, the opposite effect occurs. The force of *cohesion* between mercury molecules is greater than the force of *adhesion* between mercury molecules and glass molecules. The mercury therefore sinks down the tube to enable the mercury molecules to keep together.

The capillary action is greater for narrow tubes, because the weight of liquid having to be supported or held down is less.

Applications of capillary action. Towels, and the wicks of paraffin heaters, use capillary action. Both contain a fine network of gaps called *pores* into which liquids will soak. In towels this effect is used simply to remove water; in wicks, to transfer the liquid from the store to the flame.

Capillary attraction is often a nuisance. House bricks and concrete are also porous – capillary action is liable to draw ground water up through them, making the building damp. The photograph shows how this problem is overcome:

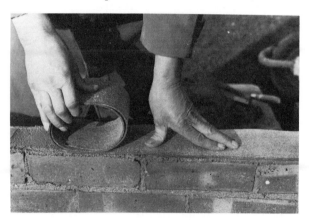

Laying a damp proof course.

A water-proof layer made from plastic is placed between a layer of bricks at the bottom of the house. This stops water being drawn up from the ground. It is called a *damp course.* There is also a gap in the middle of the wall. This prevents rain from reaching the inside of the wall.

Exercises
1. What is surface tension?
2. Explain the following observations:
(a) A steel needle can be made to float on the surface of water.
(b) Small drops of mercury form spheres.
(c) A loop of cotton floating in a soap film has an irregular shape, but when the film inside the loop is burst, it is circular.
3. Why are clothes most easily washed in warm water containing detergent?
4. Why do you think that a canvas tent is waterproof, despite the fact that there are small gaps between the cotton threads?
5. What is meant by *capillary action*?

The change of state in action

This section shows how changing conditions affect the way in which boiling and melting happen, and the practical use that can be made of these effects.

The effect of pressure on boiling

If the air pressure above some water in a flask is reduced, the liquid starts to boil at a much lower temperature than usual:

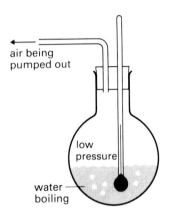

If, on the other hand, the pressure above the water is raised, the water has to be heated to a temperature above 100 °C before it will boil.

Pressure cooker. The pressure cooker is a strong aluminium pan, whose lid is sealed on with a rubber sealing ring that prevents steam escaping from inside the pan. As the water inside it is heated to boiling, steam pressure inside builds up, causing the boiling temperature to rise up to 120 °C. The high temperature makes the food cook quicker.

In order to prevent the pressure building up to a dangerous extent, the cooker is fitted with a pressure control valve. This consists of a weight which rests over a hole in the lid. When the pressure rises above a certain level it lifts the weight so that some of the steam can escape:

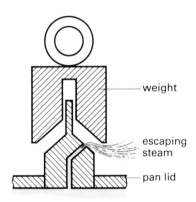

The escape of steam reduces the pressure in the pan – the weight falls back again stopping more steam from escaping. The pressure inside the pan can be increased by using heavier weights on the valve.

The refrigerator

The cooling system of a refrigerator takes heat from the ice box inside, and releases it in a series of tubes, called *condenser tubes,* mounted outside the refrigerator at the back:

24

The internal workings of the refrigerator are shown in the diagram. A fluid called freon is circulated round the pipes by a pump. The freon changes from a liquid to a vapour in the ice box, and from a vapour back to a liquid in the condenser pipes:

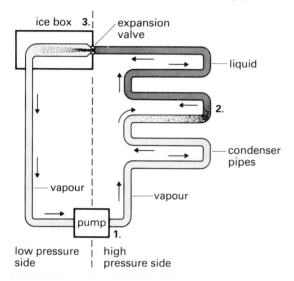

The cooling cycle has three stages:

1. Compressing. The pump takes freon vapour at a low pressure from the ice box, and compresses it against the expansion valve at the far end of the condenser pipe.

2. Condensing and giving out heat. The high pressure vapour condenses to a liquid in the condenser pipes and gives up its latent heat. The condenser pipes become hot, and are cooled by air which must be allowed to circulate round them.

3. Evaporating and cooling. Some of the freon is forced through the small hole of the expansion valve. The liquid freon, now at a low pressure, rapidly evaporates. It takes the necessary heat from the ice box and makes it very cold. The freon then circulates back to the pump for the cycle to continue.

The effect of pressure on melting

Increased pressure lowers the melting temperature of a substance. It is this effect that makes ice skating possible. The pressure of the ice skate melts the ice under it, so that there is a thin layer of water between the skate and the ice. The water acts as a lubricant, and almost completely removes the friction between the skate and the ice.

The effect of impurities on melting and boiling

Impurities affect the way in which melting and boiling occur.

Melting. When salt is added to a mixture of ice and water, the temperature drops:

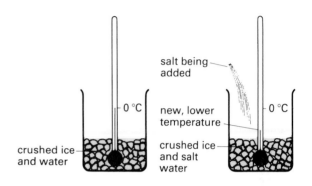

The fact that there is water present, even though the temperature is below 0 °C, shows that impurities lower the melting point. This effect is used in winter – salt spread on the roads will stop ice forming.

Boiling. When salt is added to water, the mixture must be heated to a higher temperature before it boils. Impurities raise the boiling temperature.

Exercises

1. What effect does a change of pressure have on the boiling and melting temperature?
2. Why do potatoes cook more quickly in a pressure cooker than in an open pan?
3. Draw a diagram of the cooling system of a refrigerator and explain how it works.
4. Explain why it is that the snow on the road where the traffic runs over it, is the first to melt.
5. Explain why boiling jam can cause a worse scald than boiling water.
6. In America a substance called calcium chloride is spread on the roads in winter. Explain why this is done.

Specific heat capacity

If a kilogram of aluminium and a kilogram of iron are heated at the same rate, it takes twice as long to raise the temperature of the aluminium by one degree, as it does the iron.

The aluminium needs twice as much heat as the iron. The *specific heat capacity* of aluminium is twice that of iron.

Specific heat capacity
The specific heat capacity is defined as the quantity of heat required to raise the temperature of one kilogram of the substance by one degree Kelvin. Specific heat capacity is measured in joules per kilogram per degree Kelvin (J/kg K).

The kinetic theory suggests that the specific heat capacity of a substance is a measure of the quantity of heat needed to increase the speed of its molecules a certain amount. Some are less ready to increase their speed than others – they have a higher specific heat capacity than others.

The heat formula
The actual amount of heat needed to increase the temperature of a substance will depend on how much of it there is, and the required temperature rise. The quantity of heat in joules needed to raise the temperature of an object is found from:

heat = mass × specific heat × temperature rise

$$H = m \times c \times (\theta_2 - \theta_1)$$

Where θ_1 is the initial temperature and θ_2 is the final temperature.

The same formula is used to find the heat given out when an object cools:

heat = mass × specific heat × temperature drop

$$H = m \times c \times (\theta_2 - \theta_1)$$

This formula can be used to find the specific heat of a substance experimentally. Of the various methods, two are explained here.

Specific heat capacity by a mechanical method
Some lead shot, whose temperature has been measured, is placed in a tube as shown in diagram **1**. The tube is turned upside down, keeping the lead at the same end of the tube, diagram **2**. The lead shot then falls the length of the tube, diagram **3**:

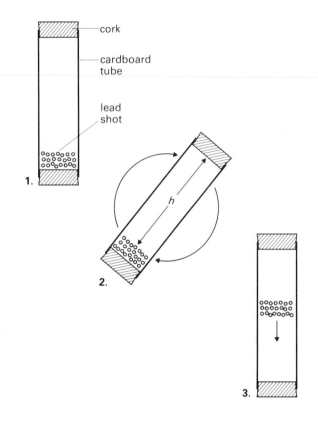

1.

2.

3.

The lead loses potential energy in falling down the tube. This energy is converted into heat energy when the lead hits the bottom of the tube. The temperature rise of the lead is measured after the lead has been made to fall down the tube N times. Then:

heat gained by lead = **potential energy lost by lead in falling N times**

But:

heat gained by lead = **mass of lead** × **specific heat** × **temperature rise**

$$= m \times c \times (\theta_2 - \theta_1)$$

potential energy lost by lead in falling N times $= N \times m\,g\,h$

Putting these together:

$$m \times c \times (\theta_2 - \theta_1) = N \times m\,g\,h$$

$$c = \frac{N \times g \times h}{(\theta_2 - \theta_1)} \quad \text{J/kg K}$$

25

Specific heat capacity by the method of mixtures

When a hot object is placed in contact with a cold object, heat flows from the hot one to the cold one until they are both at the same temperature. If there is no heat lost to the surroundings:

$$\frac{\text{heat lost}}{\text{by hot object}} = \frac{\text{heat gained}}{\text{by cold object}}$$

When this method is used experimentally, the cold object is usually water, and the hot object is the substance being investigated. For example: hot copper rivets may be heated up to 100 °C in a water bath as shown. They are then poured into water in a heat retaining vessel called a calorimeter:

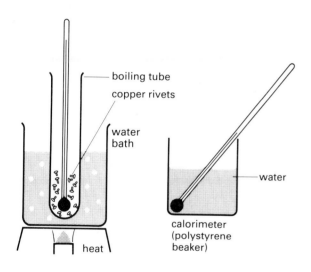

boiling tube
copper rivets
water bath
heat
water
calorimeter (polystyrene beaker)

The copper rivets cool down and the water warms up. The heat lost by the copper rivets is equal to the heat gained by the water.

The copper rivets of mass m are poured into the calorimeter which contains a mass w of water whose temperatures θ_1 has been measured. The mixture of copper rivets and water is stirred and the final temperature θ_2 is measured. By comparing the heat lost by the copper with the heat gained by the water, the specific heat capacity of the copper can be found:

$$\frac{\text{heat lost}}{\text{by copper}} = \frac{\text{heat gained}}{\text{by water}}$$

But:

$$\frac{\text{heat lost}}{\text{by copper}} = \frac{\text{mass of}}{\text{copper}} \times \frac{\text{specific}}{\text{heat of}} \times \frac{\text{temp drop}}{\text{of copper}}$$
$$\text{copper}$$

$$= m \times c \times (100 - \theta_2)$$

$$\frac{\text{heat gained}}{\text{by water}} = \frac{\text{mass of}}{\text{water}} \times \frac{\text{specific}}{\text{heat of}} \times \frac{\text{temp increase}}{\text{of water}}$$
$$\text{water}$$

$$= w \times 4200 \times (\theta_2 - \theta_1)$$

Putting these together:

$$m \times c \times (100 - \theta_2) = w \times 4200 \times (\theta_2 - \theta_1)$$

$$c = \frac{w \times 4200 \times (\theta_2 - \theta_1)}{m \times (100 - \theta_2)}$$

From this, the specific heat can be calculated. Note that this method, unlike the first one, is really only comparing the specific heat of one substance with that of water.

Exercises

1. What is meant by specific heat capacity?
2. Write down the "heat formula", stating what the symbols mean and the unit for each one.
3. What quantity of heat is needed to raise the temperature of 0.01 kg of water from 10 °C to 90 °C?
4. 0.2 kg of water cools from 70 °C to 20 °C. What quantity of heat is given out?
5. How much heat is given out when 0.3 kg of steel cools from 1023 °C to 23 °C?
6. When a mass of iron rivets were allowed to fall 100 times down a tube of length 1.92 m, the temperature of the iron was found to increase from 19 °C to 22 °C. Find the specific heat capacity of iron.
7. What is meant by "the method of mixtures"?
8. 0.5 kg of aluminium at 100 °C are dropped into a calorimeter containing 1.0 kg of water at 23 °C. If the temperature of the mixture is 30 °C, find the specific heat capacity of aluminium.

specific heat capacity of water = 4200 J/kg K
specific heat capacity of steel = 640 J/kg K
acceleration due to gravity g = 10 m/s²

Latent heat

The kinetic theory suggests that in order to change a solid into a liquid, heat energy must be provided to free the molecules from the lattice forces that hold them together.

When a liquid solidifies, the heat needed to free the molecules from the lattice forces is given out again. The heat energy difference between the solid and liquid states is called *latent heat of fusion*.

Latent heat of fusion
It is easier to demonstrate this as a liquid cools, rather than as a solid is heated. Naphthalene provides a convenient substance. It is heated to 100 °C in a water bath, and then left to cool:

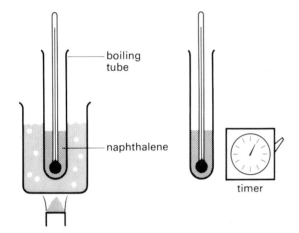

The temperature is measured every minute and a cooling curve is drawn which shows how the temperature drops as time passes:

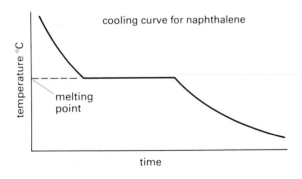

The temperature remains constant whilst the naphthalene solidifies. The temperature remains steady because during this time, latent heat of fusion is being given out. This prevents the temperature from falling.

If the naphthalene were to be heated again it would be found that the temperature would stay constant whilst the naphthalene melted. The heat being supplied would be used as latent heat to free the molecules from the lattice forces which had built up during solidifying.

Latent heat of vaporization
Latent heat is also needed to change a liquid into a gas. This is called the *latent heat of vaporization*. As boiling occurs, all the heat being supplied is used in changing the liquid to gas – boiling therefore takes place at a constant temperature. The heat is being used to free the molecules from the forces of attraction holding them in the liquid.

Specific latent heat formulas
The amount of heat needed to melt a block of 1 kg of ice is different from the amount of heat needed to boil 1 kg of water. When a substance is at its melting point, the actual amount of heat needed to melt it is given by the formula:

heat = mass × specific latent heat of fusion

$$H = m \times L_{\text{fusion}}$$

A similar formula applies to boiling – the only difference is that the specific latent heat of $L_{\text{vaporization}}$ is used.

Specific latent heat of fusion. This is the amount of heat needed to change one kilogram of a substance from solid to liquid without altering its temperature.

Specific latent heat of vaporization. This is the amount of heat needed to change one kilogram of a substance from liquid to gas without altering its temperature.

Specific latent heat is measured in joules per kilogram (J/kg); experiments to find its value for the fusion and vaporization of water are explained here.

Specific latent heat of fusion of ice
In this experiment pieces of ice are melted in warm water inside a calorimeter. The latent heat needed to melt the ice comes from cooling the warm water. Pieces of ice are added until the mass w of water is cooled from θ °C to 0 °C. The mass m of ice that was needed is found by subtracting the mass of the

calorimeter and its contents at the beginning of the experiment, from that at the end.

heat lost by warm water = **heat used in melting ice**

But:

heat lost by warm water = mass of water × specific heat of water × temp drop

$$= w \times c \times (\theta - 0)$$

heat used in melting ice = mass of ice × specific latent heat of ice

$$= m \times L_{\text{fusion}}$$

Putting these together:

$$L_{\text{fusion}} \times m = w \times c \times \theta$$

$$\therefore L_{\text{fusion}} = \frac{w \times c \times \theta}{m}$$

Specific latent heat of vaporization of water

In this experiment steam is made to condense in a thick metal calorimeter as shown in the diagram:

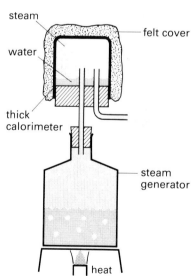

steam
felt cover
water
thick calorimeter
steam generator
heat

The latent heat given out by the steam as it condenses is used to heat up the calorimeter.

The mass w of the calorimeter, and its initial temperature θ °C is measured. The calorimeter is placed on the steam generator and left until steam comes out of the side tube. The calorimeter is now at 100 °C. The mass m of the condensed steam is

found by subtracting the mass of the calorimeter, from the mass of the calorimeter and condensed steam.

heat given out by steam condensing = **heat gained by calorimeter**

But:

heat given out by steam condensing = mass of steam × specific latent heat

$$= m \times L_{\text{vaporization}}$$

heat gained by calorimeter = mass of calorimeter × its specific heat capacity × temp rise

$$= w \times c \times (100 - \theta)$$

Putting these together:

$$m \times L_{\text{vaporization}} = w \times c \times (100 - \theta)$$

$$\therefore L_{\text{vaporization}} = \frac{w \times c \times (100 - \theta)}{m}$$

Exercises

1. Fill in the sentences using the words:
The * of a * remains * whilst it is * state.
The * needed to * or * a substance is called the * heat. (constant, changing, temperature, substance, latent, heat, vaporize, melt)
2. Write down the latent heat formula. State what the symbols mean and what its units are.
3. What is the quantity of heat needed to melt 0.1 kg of ice at 0 °C? The specific latent heat of fusion of ice is 336 000 J/kg.
4. What is the quantity of heat needed to vaporize 0.01 kg of boiling water? The specific latent heat of vaporization of water is 2 260 000 J/kg.
5. How much heat is given out when 0.2 kg of aluminium at its melting point solidifies? Its specific latent heat of fusion is 170 000 J/kg.
6. 0.2 kg of ice are stirred into 1.0 kg of water at 18 °C in a calorimeter. Melting this ice was just enough to reduce the water's temperature to 0 °C. Find the specific latent heat of fusion of ice.
7. Steam was passed into a copper calorimeter of mass 1.0 kg initially at 28 °C until it was at 100 °C. It was found that 0.012 kg of steam had condensed. If the specific heat capacity of copper is 385 J/kg K find the specific latent heat of vaporization of water.

The gas laws

The pressure, volume and temperature of a gas all affect one another. Change one and you automatically change the other two. This would make the results of an experiment to investigate changes in all three at once complicated to understand. This problem is overcome by making one of them stay constant, whilst the relation between the other two is investigated.

Boyle's law
This is about the variation of volume with pressure, of a gas at steady temperature. The apparatus shown in the diagram below may be used to find how the volume of a fixed mass of gas varies with pressure, at constant room temperature:

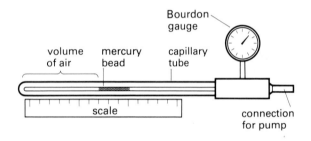

A bicycle pump is used to increase the pressure which is measured on the bourdon gauge. The fixed mass of air is trapped in the capillary tube by a bead of mercury, and its volume is measured on a scale.

When the pressure goes up, the volume goes down so that:

pressure × volume = constant
$$P \times V = \text{constant}$$

Boyle's law states that:

For a particular mass of gas, the pressure multiplied by the volume stays constant, provided that the temperature does not change.

Kinetic theory of Boyle's law. The pressure of a gas is caused by the molecules hitting the walls of the container. Reducing the volume forces the molecules closer together. So the number of molecules hitting the walls increases – the pressure increases.

The pressure law
This is about the variation of pressure with temperature at constant volume.

The apparatus shown in the diagram below may be used to find how the pressure of a fixed volume of air varies with temperature:

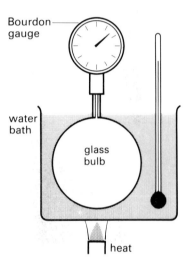

The air is contained in a glass bulb whose volume is constant, and the pressure is measured on the bourdon gauge. A water bath is used to vary the temperature. The temperature must be measured in degrees kelvin (the absolute temperature). When the temperature goes up, the pressure goes up. The pressure increases in step with the absolute temperature – the pressure *is proportional to* the absolute temperature. This can be expressed mathematically as:

$$\frac{\text{pressure}}{\text{absolute temperature}} = \text{constant}$$
$$\frac{P}{T} = \text{constant}$$

The pressure law states that:

For a particular mass of gas the pressure is proportional to the absolute temperature, provided that the volume does not change.

If the absolute temperature of a gas increases the speed of the molecules increases. The force of the impacts on the walls of the container increases, so the pressure increases.

27

Charles' law

This is about the variation of volume with temperature at constant pressure. The apparatus shown in the diagram below is used to find how the volume of some air varies with its temperature:

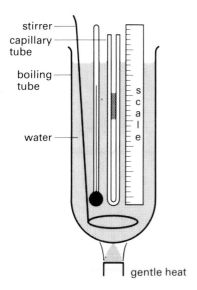

The pressure stays steady, since it is equal to atmospheric pressure plus the small pressure due to the weight of the bead of mercury trapping the air. The volume of air is measured on the scale and the water bath is used to vary the temperature. The absolute temperature must again be used.

It is found that the volume of the gas increases in step with the absolute temperature – the volume *is proportional* to the absolute temperature. This can be expressed mathematically as:

$$\frac{\text{volume}}{\text{absolute temperature}} = \text{constant}$$

$$\frac{V}{T} = \text{constant}$$

Charles' law states that:

For a particular mass of gas, the volume is proportional to the absolute temperature, provided that the pressure does not change.

If the absolute temperature of a gas increases, the speed of the molecules increases. For the pressure to remain steady there must be less molecules hitting each part of the walls so the volume must increase.

The general gas law

The three gas laws can be summarised by one equation:

$$\frac{\text{pressure} \times \text{volume}}{\text{absolute temperature}} = \text{constant}$$

$$\frac{P \times V}{T} = \text{constant}$$

the general gas law equation is frequently written in the form:

$$\frac{P_1 V_1}{T_1} = \frac{P_2 V_2}{T_2}$$

Where P_1, V_1 and T_1 refer to one set of conditions of pressure, volume, and temperature, and P_2, V_2 and T_2 to another set of conditions.

An individual gas law can be obtained from this equation by covering up the variable which is kept constant.

Exercises

1. A mass of gas at a pressure of 20 N/m² has a volume of 3 m³. What will be the volume if the pressure is doubled, at constant temperature?
2. The volume of a mass of gas is reduced from 5 m³ to 2 m³. If the pressure was initially 40 N/m² what will be the new pressure if the temperature remains constant?
3. The pressure of a fixed volume of gas at 200 K is 50 N/m². What would be the pressure if the temperature was increased to 300 K?
4. If pressure of a fixed volume of gas at 300 K is increased from 5 N/m² to 10 N/m², what will the new temperature be?
5. The temperature of 6 m³ of gas is increased from 300 K to 400 K. What will be the new volume of the gas if the pressure remains constant?
6. The volume of a gas is increased from 10 m³ to 20 m³ at constant pressure. If the initial temperature was 300 K find the new temperature.
7. A mass of gas has a volume of 5 m³, a pressure of 20 N/m² and a temperature of 300 K. What will be the new pressure if the volume is changed to 4 m³ and the temperature to 400 K?

Questions on chapter 4

1. The two diagrams below represent the arrangement of the molecules in a solid and a liquid:

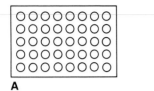
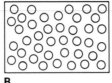

A   **B**

  a Write down which diagram represents:
    i A solid
    ii A liquid
    iii Give a reason for your answer.
  b Draw a similar diagram to represent the molecules of a gas.
  c When solids, liquids or gases are heated what happens to:
    i The speed of the molecules?
    ii The distance apart of the molecules?
    iii The density of the material?
  d Explain in terms of the movement of the molecules why a gas exerts pressure. (M.R.E.B)

2. **i** State two differences between evaporation and boiling.
  ii What effect does an increase in pressure have on the boiling point of a liquid?
  iii State one practical use of this effect.
                                  (W.J.E.C)

3.

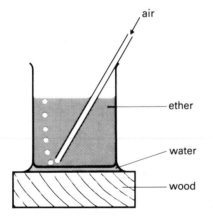

In the experiment shown above, what happens to the water if air is blown through the ether for a sufficiently long period? Why?
  Air is blown through the ether in order to increase the rate of evaporation. State one other way by which the rate of evaporation by a liquid may be increased.   (W.M.E.B.)

4. A few grams of a substance called salol were placed in a test tube with a thermometer. The tube was heated in water bath and when the salol had melted, the tube was taken from the bath and the temperature recorded every minute. The results obtained are given below.

time (min)	temp °c	time (min)	temp °c
0	61·0	7	42·0
1	53·0	8	42·0
2	48·5	9	41·7
3	45·2	10	40·8
4	43·2	11	39·5
5	42·3	12	38·5
6	42·0		

Use the results to plot a cooling curve (on graph paper).
From the curve find the melting point of salol and explain the shape of the curve.   (A.L.S.E.B.)

5.

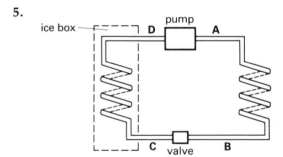

The diagram shows a simple compression type of refrigerator.
  i By referring to the behaviour and condition of the liquid or vapour in the regions marked **A**, **B**, **C** and **D**, explain how the system operates.
  ii Why is it impossible to cool down the kitchen by leaving the refrigerator door open, while the refrigerator is operating?   (W.M.E.B.)

6. **a** Copy the diagram into your book drawing in the water levels in the two glass tubes.

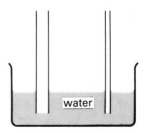

b The "damp-proof course" of a house is a non-porous layer between two rows of bricks. Why is it desirable? (E.A.E.B.)

7. a 100 g of water at 80 °C is poured into a copper vessel at 20 °C and it is found that the temperature falls to 45 °C.
   i What is the rise in temperature of the vessel?
   ii Explain what has happened to the heat energy lost by the water.
  b The same experiment in **a** is repeated using 100 g of paraffin at 80 °C in place of water. The temperature rise of the vessel is much less than that produced in experiment **a**. How do you explain this? (W.J.E.C.)

8. a What is meant by "specific heat capacity"?
  b Describe an experiment to determine the specific heat capacity of a solid.
  c A copper block of mass 3 kg is heated in an oven. Its temperature rises 100 °C in 2 minutes.
   i Taking the specific heat capacity of copper as 400 J kg^{-1} K^{-1}, work out how much heat the block has absorbed.
   ii At what rate was the block absorbing heat?
   iii If only 50% of the heat generated by the oven was absorbed by the block, at what rate was the oven generating heat? (E.A.E.B.)

9. Calculate the heat given out by 200 g of a molten substance, at its melting point, as it solidifies without further change of temperature. Assume that the specific latent heat of the substance is 150 J/kg. (N.W.R.E.B.)

10. a The following piece of apparatus was used to investigate the pressure and volume of a gas.

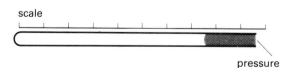

   i What happens to the volume of trapped air as the pump increases the pressure?
   ii What care must be taken when reading the oil level?

b In a Boyle's law experiment the readings were taken as follows:

Vol (cm³)	30	27	22	18	14
Pressure (kN m^{-2})	145	175	220	260	300

   i Plot a graph of pressure (Y-axis) against volume (X-axis).
   ii From the graph predict the value of the volume if the pressure is 280 kN m^{-2}.
   iii From the graph what is the value of the pressure if the volume is 26 cm²?
  c Why must the temperature be kept constant during the experiment?
  d How could the results be used to verify Boyle's law? (A.L.S.E.B.)

11. Lead shot may be heated up if it is put in a long cardboard tube which is repeatedly turned upside down. Assume that the tube contains 1 kg of lead shot and that each time the tube is inverted the shot falls 1 metre.

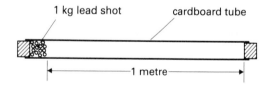

   i What force is need to lift the 1 kg mass of lead? (g = 10 m/s²)
   ii How much work is done when the tube is inverted once?
   iii How much work is done when the tube is inverted 100 times?
   iv Why does inverting the tube cause a temperature rise of the lead shot?
   v If the temperature rise caused by 100 inversions is 6°C and the specific heat capacity of lead is 130 J/kg K, how much heat is gained by the lead shot?
   vi Comparing the heat gained with the work put in, what does the efficiency of this energy conversion appear to be?
(M.R.E.B.)

Expansion of solids

When a solid is heated it becomes bigger – it *expands*. The amount of expansion is fairly small. Expansion can be demonstrated by means of the apparatus shown below:

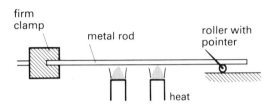

The heated rod expands and rolls the needle, which moves the pointer. When the heat is removed the metal *contracts* back to its original length. All substances expand by different amounts. Even different metals expand by different amounts.

The fact that substances expand as they get hotter can be useful, but more often expansion is a nuisance, and allowances have to be made for it.

Problems caused by expansion
Expansion can cause a number of problems:

Bridges. The part of a bridge which spans a road, railway or river is frequently mounted on rollers, to allow it to expand on hot days, and to contract on cold ones. The bridges which span motorways show many of the ways in which the problem has been solved. The photos show a "road view" and a "side view" of expansion joints:

Railway lines. The distortion of railway lines due to expansion in hot weather can be prevented by the use of overlapping joints. As the rails expand, the increased length is taken up by the two rails sliding past one another, as shown in the photograph below. Such joints as these are only required occasionally, since the most modern method of allowing for the expansion of railway lines is to weld them together and then stretch them, so that any expansion only results in slightly reduced tension.

Oil pipe-lines. The desert regions where oil is found, are very hot in the day and cold at night. Pipelines, therefore, are laid with built in zig-zags. As the pipe expands during the day, the bends increase slightly to take up the increased length. At night, the pipes become much cooler, and the zig-zags straighten out again:

28

Putting expansion to work

A *bimetallic strip* is a metal strip made of two different metals, such as copper and iron, bonded tightly together. When a bimetallic strip is heated the copper expands more than the iron so the strip bends:

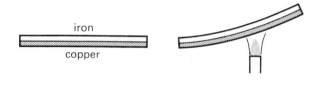

iron

copper

The thermostat

When a bimetallic strip is used to maintain a steady temperature, it is called a *thermostat*. It controls the temperature by switching the heat source on and off. The room thermostat provides a typical example. When the room is cold, the bimetallic strip allows a current to flow through the heater – the electrical contacts are closed as in diagram **1**:

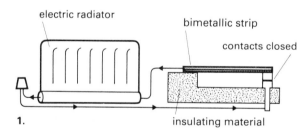

electric radiator

bimetallic strip

contacts closed

1.

insulating material

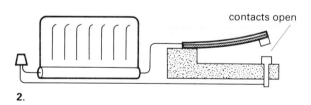

contacts open

2.

As the temperature of the room rises the bimetallic strip starts to bend up. The contacts eventually open, switching the heater off, as in diagram **2**. The room begins to cool down, so the bimetallic strip straightens again. The contacts close again, switching the heater on.

Adjustable thermostat. It is often necessary to be able to alter the temperature at which the thermostat switches on and off. For a low temperature setting, the temperature knob is screwed out as in diagram **1** below. The bimetallic strip only has to bend slightly to open the contacts. The temperature will not rise very high before the contacts open:

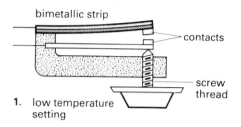

bimetallic strip

contacts

screw thread

1. low temperature setting

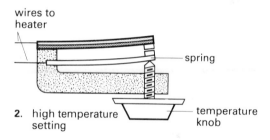

wires to heater

spring

2. high temperature setting

temperature knob

When the temperature knob is screwed in as shown in diagram **2**, the lower contact is pushed up. The bimetallic strip now has to bend much further before the contacts open. A higher temperature must be reached before the contacts open.

Exercises

1. Explain, with the aid of a diagram, how the problem of expansion is overcome in railway lines, bridges and oil pipelines.

2. How could you remove a metal screw cap which is tightly fixed on a glass jar?

3. Copy out the following sentences filling in the missing words from the list.

Brass expands more than * so a * strip made of the * metals bends when *. When allowed to cool, the metals * and the strip becomes * again. (bimetallic, contract, heated, steel, two, straight).

4. Explain with the aid of a diagram how an automatic fire alarm using a bimetallic strip works.

Expansion of fluids

When liquids and gases get hot they expand, just as solids do. When heated by the same amount, liquids expand about ten times more than solids, and gases expand about a thousand times more than liquids.

Expansion of liquids

The simple experiment below demonstrates the expansion of water. When the flask is heated the water expands and rises up the tube:

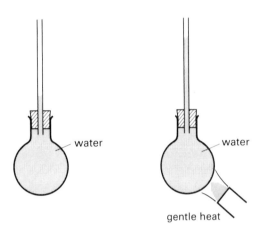

As with solids, different liquids expand by different amounts.

Expansion of gases

Unlike liquids and solids, all gases expand at the *same* rate! The expansion of air is demonstrated as shown in the diagram below:

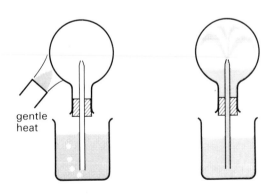

The air expands as the flask is heated. Bubbles of air are forced out through the water in the beaker.

When the flask is allowed to cool the air contracts. Water is forced into the flask in a fountain – it replaces the air that has been forced out by heating.

Hot air balloon. When a substance expands, it becomes less dense. This effect is used to make hot air balloons. The air in the balloon is heated by a gas burner, so that it expands and becomes less dense than the surrounding air. This makes the balloon rise:

The theory of hot air balloons is quite simple. The practice is sometimes harder . . .

Water and ice

The behaviour of water at and around its freezing point is very strange. Firstly, when a certain volume of water freezes the volume of ice formed is nearly one tenth greater than the volume of the water. It is this increase in volume of water on freezing, which causes burst water pipes.

29

Icebergs – off the coast of Iceland, strangely enough.

The peculiar expansion of water

As hot water cools it contracts – it becomes *more* dense – until a temperature of 4 °C is reached. But when water cools further from 4 °C to 0 °C, it expands and becomes slightly *less* dense than it was at 4 °C. The graph below shows how the density of a mass of water varies with temperature:

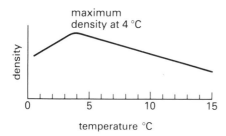

Freezing of a pond. This peculiar expansion of water determines the way in which a pond freezes over. When the air temperature is below 0 °C and the water temperature is above 4 °C, the cold air cools the water at the surface. The cooler water becomes more dense and sinks to the bottom, as warm water from the bottom rises to take its place:

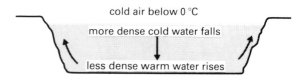

Because of this movement, the temperature throughout the water drops steadily. After all the water has cooled to 4 °C, the top layers *expand* as they get colder. The top layer becomes less dense than the warmer water below, so the colder water stays at the top:

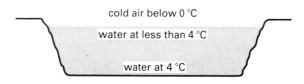

Eventually a layer of ice forms at the top and since ice is less dense than water, it stays there. Ice is a poor conductor of heat so the rest of the water in the pond cools at a much slower rate. Fishes and plants can live quite comfortably in the water at 4 °C under the ice.

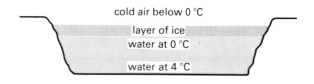

Exercises

1. Describe an experiment to demonstrate (a) the expansion of a liquid and (b) the expansion of a gas.
2. Explain the method by which a hot air balloon rises.
3. A pond has a layer of ice over its surface. What are the temperatures of the water just underneath the ice and near the bottom of the pond? How does the layer of ice slow down the complete freezing of the water in the pond?

Thermometers

Some things are hot and some things are cold – the *temperature* of an object indicates how hot it is. A thermometer is used to measure temperature.

The liquid in glass thermometer
The diagram below shows a typical thermometer, which may contain alcohol or mercury. These, unlike water, expand at a steady rate and do not freeze as easily:

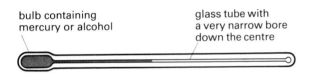

bulb containing mercury or alcohol

glass tube with a very narrow bore down the centre

When the bulb is heated the liquid expands and forces its way along the bore of the tube. The higher the temperature the more the expansion; the position that the liquid reaches is a measure of the temperature.

Temperature is measured in degrees Celsius. This scale of temperature is made by using two easily obtained temperatures, called *fixed points,* as reference points. The distance between them is then divided into one hundred parts, or *degrees.*

Lower fixed point. The temperature of pure melting ice is taken as 0 °C:

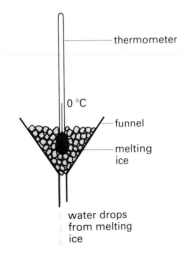

thermometer

0 °C

funnel

melting ice

water drops from melting ice

Upper fixed point. The temperature of steam from pure boiling water under standard atmospheric pressure, is taken as 100 °C:

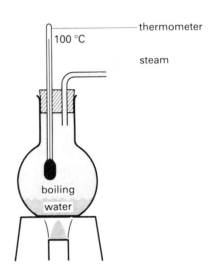

thermometer

100 °C

steam

boiling water

The position of the liquid at 0 °C and 100 °C is marked on the glass of the thermometer using the two pieces of apparatus shown above. The distance between these two marks is then divided into a hundred parts to make the Celsius scale of temperature.

Special thermometers
The thermometer described above is a general purpose thermometer but special thermometers are needed to do particular jobs.

Clinical thermometer. This thermometer is used to measure body temperature. It has a very fine bore tube to make it sensitive. A small expansion of the mercury makes it shoot a long way along the tube. The glass of the bulb is also very thin so that body heat can get to the mercury quickly to make the thermometer quick acting:

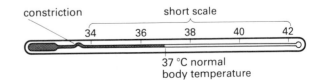

constriction

short scale

34   36   38   40   42

37 °C normal body temperature

30

When the thermometer bulb is placed in the mouth, the mercury expands and is forced past the constriction, along the tube. When the thermometer is removed from the mouth, the bulb cools and the mercury in it contracts very quickly. The mercury column breaks at the constriction, leaving the mercury in the tube in place:

This gives the nurse time to accurately read the patient's body temperature. The thermometer is reset by shaking the mercury back into the bulb.

Maximum and minimum thermometer. This type of thermometer is used by gardeners and weather men to measure the highest and lowest temperatures during the day. It has alcohol, mercury, and *riders* inside it:

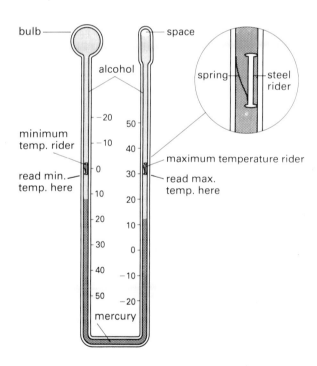

The temperature is measured by the alcohol in the bulb. The amount of mercury is so small that its expansion may be ignored. The alcohol pushes the mercury in the tube, and the mercury pushes the riders. The alcohol will *not* push the riders, but flows past them. The diagrams at the top of the next column show how the thermometer works:

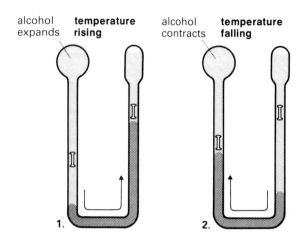

Diagram **1** shows what happens when the temperature rises. The alcohol in the bulb expands, flows past the minimum rider and pushes the mercury round.

The mercury in turn pushes the maximum rider along as the temperature rises. The mercury-end of the rider marks the highest temperature reached.

When the temperature falls, as in diagram **2**, the alcohol contracts pulling the mercury back. The maximum rider is left where it was, and the minimum rider is pushed by the mercury into the minimum position. The riders may be reset by moving them back to the mercury with a magnet.

Exercises
1. Draw a diagram of a clinical thermometer and explain how it differs from an ordinary one.
2. Why would it be unwise to sterilize a clinical thermometer in boiling water?
3. What are the highest, lowest, and actual temperatures recorded on the maximum and minimum thermometer in the diagram?
4. Give two reasons why water is not a good liquid to use in thermometers for measuring weather temperatures.
5. Draw a diagram of a maximum and minimum thermometer and explain how it works. What is this type of thermometer used for?

Transmission of heat — conduction

Heat flows from hot things to cold things. The flow of heat through a substance from hot to cold parts of it is called the *conduction of heat:*

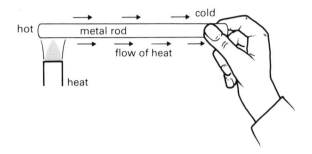

Different substances conduct heat at different rates. Their *thermal conductivities* are different. For example, copper is a very good conductor of heat; it has a high thermal conductivity. Wood is a bad conductor of heat; it has a low thermal conductivity. Generally metals are good conductors of heat. Liquids, gases and non-metals such as wood, polystyrene and glass are poor conductors of heat. A substance that is a poor conductor of heat is called an *insulator*.

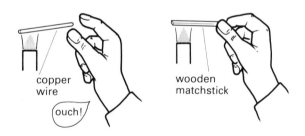

Conduction and insulation may be simply explained using the kinetic theory. Hot places are ones where there is more vibration of the molecules; heat conducts by impact of the hot molecules with the cooler ones next to them. Conductors transfer this vibration easily; insulators do not.

Using conduction of heat

Often, the conduction of heat causes problems and attempts are made to prevent it. Sometimes, the conduction of heat is useful, so the designer or engineer will choose a material that has a high thermal conductivity.

Saucepans: transfer of heat. The pan is frequently made of aluminium. Aluminium is a good conductor of heat; heat is quickly conducted from the heat source to the contents of the pan.

Transistors. Electrical components such as transistors sometimes get hot in operation. When this is the case, they are often mounted in metal fins that conduct the heat away from them.

Preventing conduction of heat

More often the problem is how to prevent conduction rather than how to aid it. The following sections show how this is done.

Clothing. Air is a bad conductor of heat. All clothing works on the principle that by trapping layers of air next to the skin, heat loss is prevented. The colder the weather, the thicker the clothing – the thicker the layer of air. A string vest, being mainly air space, is efficient at keeping its wearer warm.

Hot water tank insulation. All hot water tanks should be covered with a quilted fibreglass jacket. Heat from the hot water can only conduct slowly through the cover – this keeps the water hot.

Insulating a house. Heat costs money, so it is important to reduce the amount of heat lost. Heat is lost through the roof, through the walls, and through the windows.

Roof insulation. Rolls of fibreglass roof insulation are laid across the floor of the loft. This reduces the amount of heat lost through the roof:

31

Wall insulation. The outer wall of a house is made of two sets of bricks with a gap between them called a *cavity*. The loss of heat through the walls is much reduced if the cavity is filled with insulating material:

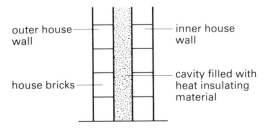

outer house wall — inner house wall

house bricks — cavity filled with heat insulating material

Double glazing. The heat lost through the glass of a window is reduced by having two panes of glass separated by a thin sandwich of air. Air is a bad conductor of heat:

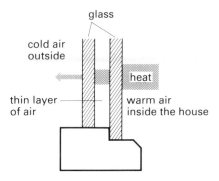

glass

cold air outside

heat

thin layer of air — warm air inside the house

Laboratory experiments on thermal conductivity

Comparing the thermal conductivity of metals.
The apparatus for this is shown in the diagram below:

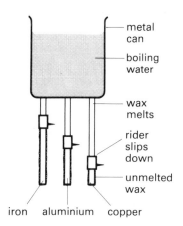

metal can

boiling water

wax melts

rider slips down

unmelted wax

iron   aluminium   copper

Heat is conducted down the rods from the boiling water. The heat is conducted quicker down the rods with the higher thermal conductivity. This melts the wax quicker and allows the rider to slide down faster.

Thermal conductivity of water. Water is a very poor conductor of heat. In the following experiment, the water at the top of the tube can be made to boil vigorously whilst ice still remains unmelted at the bottom of it:

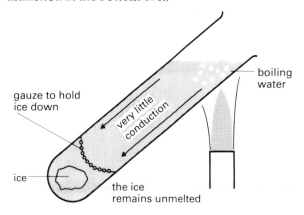

gauze to hold ice down

very little conduction

boiling water

ice

the ice remains unmelted

The water is heated at the top of the tube and the ice is placed at the bottom to prevent the heat from the bunsen being moved by *convection*. Convection of heat is explained in the next section.

Exercises
1. What is meant by conduction of heat?
2. State whether the following substances are good or bad conductors of heat.
(a) iron, (b) concrete, (c) glass, (d) copper, (e) water, (f) paraffin, (g) air, (h) brass, (i) wood.
3. Explain how a string vest keeps you warm.
4. In what three ways is heat lost from a house? Explain how these heat losses are reduced.
5. Describe an experiment to compare the thermal conductivities of brass and steel. Include a diagram of the apparatus that you would use.
6. Explain why a carpet feels warmer to the bare feet, than lino laid in the same room.

Transmission of heat — convection

Convection is a method of heat flow that only takes place in fluids. Heat flows through the fluid by means of the movement of warm parts of the fluid.

pipe is a safety precaution. If the water boils the expansion pipe shoots steam into the header tank.

Convection in liquids

Convection in water is demonstrated by the simple experiment shown below:

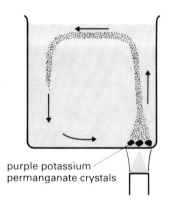

purple potassium permanganate crystals

As the heated water expands, it becomes less dense and rises. Cold, more dense water, falls to take its place. These movements, which are shown up by the purple colour of the potassium permanganate, are called *convection currents.*

Domestic hot water system. The household hot water system shown at the top of the next column makes use of convection.

The boiler heats the water; the water rises by convection up to the hot water storage tank. Cooler water in the storage tank sinks back down to the boiler. In this way a supply of hot water collects in the storage tank.

The pressure needed to force hot water out of the taps and to replace the hot water run off, is provided by a header tank in the roof. A valve called a ball-cock controls the supply of water into the tank. If the water level drops, the ballcock opens; as the water level rises, it shuts again. The expansion

Convection in gases

Convection in gases such as air happens in exactly the same way as convection in water. It may be demonstrated by means of the apparatus shown below. Smoke is used to make the convection currents visible:

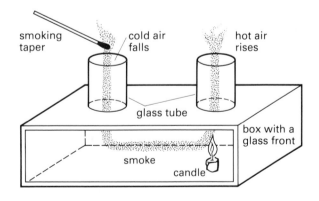

Convection of air is used in heating systems; sea breezes are also caused by convection.

32

Electric convector heater. The electric element heats the air, which rises out of the top of the case. Cold air flows in at the bottom to replace the hot air:

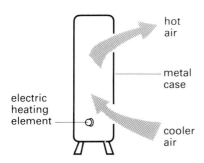

Central heating radiator. This consists of a series of pipes containing hot water. The radiator is wrongly named – it should really be called a convector. It is placed near the floor, where it heats the air. The warm air that it produces then circulates round the room in convection currents as shown:

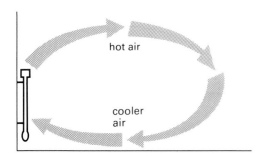

Sea breezes. On hot days, a breeze often blows from the sea towards the land. This is caused by the land heating up more quickly than the sea. The land heats the air above it, which expands, becomes less dense and rises. Cooler air from the sea then blows in to take the place of the warmer air rising above the land, as shown in diagram **1** in the next column.

At night time the breeze blows in the opposite direction, from the land to the sea. The sea cools down more slowly than the land and so is warmer than the land. The sea now heats the air above it, so that the air expands and rises. Cooler air from the land then blows in to take the place of the warmer air rising above the sea, as shown in diagram **2**.

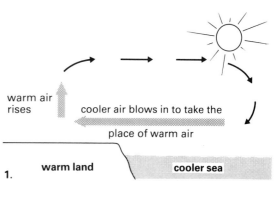

1.

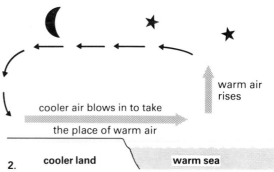

2.

Exercises
Copy out the sentences in (1) and (2) filling in the missing words from the lists.

1. Convection is the * of * through a * by the * of the fluid itself, in the * of *.
(fluid, form, flow, currents, movement, heat)

2. On a sunny day the * heats up more * than the sea. The air above the land is heated so that it *, becomes less * and *. * air blows in from the sea to take its place
(cool, dense, expands, land, quickly, rises)

3. Draw a diagram of the domestic hot water system. Explain how hot water is collected in the storage tank, how the pressure to force the hot water out of the tap is produced, and how hot water that has been drawn off is replaced.

4. Why are hot water 'radiators' placed near the floor of a room, and the cold pipes of a refrigerator placed near the top of the cabinet?

Transmission of heat — radiation

The transfer of heat by radiation is different from conduction and convection because heat radiation will travel through empty space. Heat radiation is in some ways similar to radio and light waves and is often called *infra-red* radiation.

All objects give off heat radiation – but the hotter they are, the more heat they give off. The surface temperature of the sun is over 5000 °C, so it gives off a lot of heat radiation. Heat radiation cannot be detected until it falls upon an object. The heat radiation from the sun makes the Earth hot.

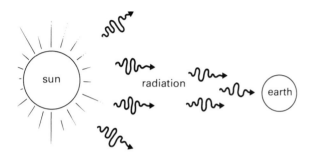

Emission and absorption of radiation
All objects continually give off radiation – which means that all objects continually receive it, too.

Emission. The rate at which an object gives out, or *emits,* heat depends not only on the temperature, but also on the nature of its surface. This is demonstrated by means of two similar cans filled with boiling water:

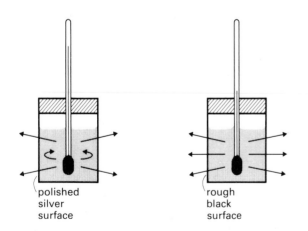

polished silver surface　　　rough black surface

As the cans are allowed to cool, it is found that the can with the rough black surface cools quicker than the can with the polished silver surface. The rough

black surface emits heat radiation faster than the polished silver surface, so it becomes cool more quickly.

Absorption of heat radiation. The same two cans may be filled with equal volumes of cold water and placed the same distance away from a radiant heat source:

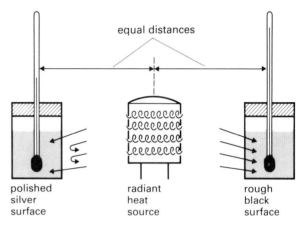

equal distances

polished silver surface　　　radiant heat source　　　rough black surface

It is found that the rough black can heats up quicker than the polished silver one. The rough black can is a better absorber of heat radiation than the polished silver one.

Law of radiation. These two experiments show that the rough black can is both a good emitter and a good absorber of radiation. The polished silver can is similarly both a bad emitter and a bad absorber. These results are summarised by the law of radiation which states:

A good emitter of heat radiation is a good absorber, and a bad emitter is a bad absorber.

Detection of heat radiation
There are various methods of detecting heat radiation:

Photography. Films can be made that are sensitive to heat radiation. In photographs using these films, warm objects appear bright and cold objects appear dark. The photograph at the top of the next page was taken in complete darkness with an infra-red film. The outline of a person on the left hand side of the photograph is very clearly visible. A "heat stain" left on the carpet after a warm body has been removed, can also be seen.

33

On a clear night heat from the ground is lost by radiation into space, so the night is cool, as in diagram **1**. On a cloudy night, heat radiation from the ground is reflected back by the clouds, so the night is warm, as in diagram **2**:

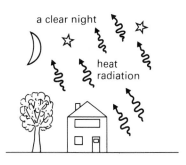

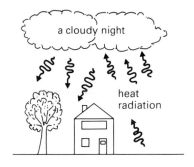

Thermocouple. This consists of two wires of different metals joined at one end. The other ends are connected to a sensitive meter:

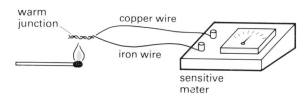

When the junction of the two wires is heated, a current flows round the circuit. The current is registered on the sensitive meter. This effect is called the thermoelectric effect.

Thermopile. A thermopile is made by connecting about forty thermocouples together to increase the current produced. Frequently a metal cone is placed over the junctions as shown, so that the thermopile only receives radiation coming directly towards it:

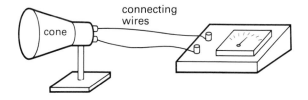

Radiation and weather

Clear nights are generally colder than cloudy nights, when other weather conditions are the same. A clear night in winter is a sign that there is going to be a frost.

Exercises
1. Explain what is meant by heat radiation.
2. Explain how an infra-red photograph of a hot black metal rod would differ from an ordinary one.
3. A metal can with one side painted white and the other side painted black is filled with boiling water. Explain how you would use a thermopile to find which surface was the better emitter of heat.
4. Explain why the weather is usually warmer when the sky is clear than when the sky is cloudy.
5. If a leaf is placed on fresh snow, after a few days of sunshine it will have sunk into the snow. Explain this fact.

Applications of heat transfer

This section gives some examples that show how understanding of the movement of heat is used.

Forced convection

The natural flow of heat by convection currents is not fast enough for many applications. The convection currents can be increased by using a fan or a pump to move the hot fluid. This is called *forced convection,* of which three examples are given here.

Hair drier. The fan draws in air at the side and blows it over the electric heating element. In this way a strong forced convection current of hot air is produced:

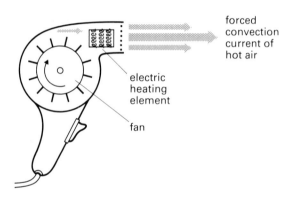

forced
convection
current of
hot air

electric
heating
element

fan

Domestic central heating. The hot water flows in and out of household radiators through narrow pipes. Hot water from the boiler will not circulate very easily by natural convection. A pump is used to force the hot water round the system:

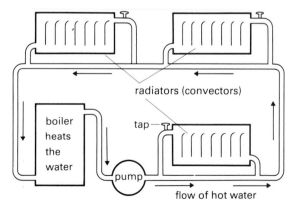

radiators (convectors)

boiler
heats
the
water

tap

pump

flow of hot water

Cooling system of a car engine. In burning its fuel, a car engine produces a lot of heat, but the engine itself has to be kept reasonably cool. This is done by circulating water through it. A pump forces currents of cool water round the engine, where the water takes up some of the heat. The hot water then passes down through the radiator which consists of a lot of narrow pipes. A fan and the movement of the car, draws cold air across the pipes of the radiator. This cools the water, which is then circulated back to the engine:

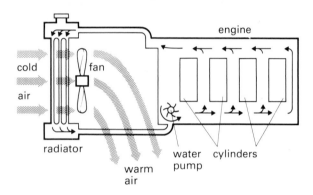

cold

air

fan

engine

radiator

warm
air

water
pump

cylinders

In older designs of car the fan is driven by the engine and is always working. When the car is going fast, this wastes power and fuel, and is noisy. The car's motion forces air through the radiator without needing a fan. In more modern cars, the fan is driven by an electric motor. The motor is controlled by a thermostat which switches the fan off when the car is going fast.

The vacuum flask

The vacuum flask is a container that can be used for keeping things hot or cold. The diagram at the top of the next page shows a vacuum flask being used to keep a liquid hot.

It is specially designed to reduce the flow of heat through its walls to almost nothing:

Radiation is reduced by silvering the inside surfaces of the glass.

Convection is stopped by having a vacuum between the glass walls.

Conduction is also stopped by the vacuum. Very little heat is conducted through the stopper as it is made of plastic which is a bad conductor.

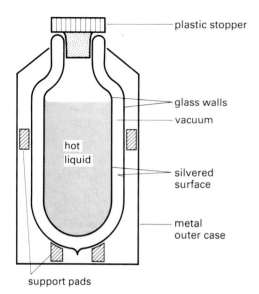

plastic stopper

glass walls

vacuum

hot liquid

silvered surface

metal outer case

support pads

The glass vessel of the vacuum flask is very delicate and can easily be broken. The glass flask is therefore fitted into a metal outer case with cushioned support pads. This enables the flask to withstand all but the most severe knocks.

The miner's safety lamp

Pockets of an explosive gas called fire-damp are found in coal mines. Any unprotected flame or spark would explode the gas. Early in the nineteenth century the miner's safety lamp was designed to overcome this problem:

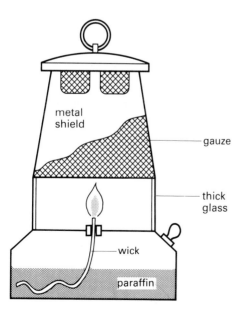

metal shield

gauze

thick glass

wick

paraffin

The metal gauze conducts away the heat of the flame so that the temperature outside the lamp never rises high enough to explode the gas. If an explosive gas enters the lamp through the gauze the flame changes colour but the gas outside cannot explode. The safety lamp is no longer used for lighting mines, but it is still used to detect explosive gases.

Exercises
1. Explain what is meant by forced convection.
2. Describe two everyday examples of forced convection.
3. Draw a labelled diagram of a vacuum flask.
4. Copy out the following passage, filling in the missing words from the list.

> The * between the two glass walls of a vacuum flask filled with ice-cream will prevent heat entering it by * and *. The inner surfaces of the walls are * to reduce the heat reaching the ice-cream by *. The plastic stopper is an * so the heat * through it is very small.

(conduction, conducted, convection, insulator, radiation, silvered, vacuum).

5. Why do mines still have paraffin safety lamps in them, when electric lamps are so much brighter?
6. The photograph below shows a hot-water tank. In what ways are heat losses prevented? Will it be as efficient as a vacuum flask?

34

Rockets, jets, and turbines

The engines described in the next three sections are all *heat engines*. In outward appearance and design they look very different. However, all these engines are driven by expanding gases, and heat energy is used to make the gases expand.

The rocket engine

A photograph, and a simplified cross section of a rocket engine of the type used in the Apollo spacecraft, are shown below:

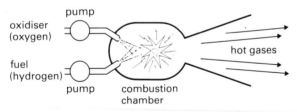

Fuel and oxidiser are pumped into the combustion chamber where they mix and explode. The explosion produces hot gases which rush out of the exhaust at a high velocity. It is the reaction to the force pushing the gases out which drives the rocket engine forward.

Rocket engines are used in space craft like the space shuttle. The rocket is the only heat engine which will work in the vacuum of space. This is because the space craft carries not only fuel, but also the oxygen to burn it.

The jet engine

The jet engine works on the same principle as the rocket engine. The main difference is that air is used to burn the fuel. The photograph and diagram show a side view of a jet engine:

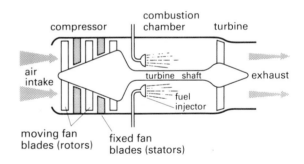

The air is drawn in to the engine by means of a series of fans called a compressor. The compressor consists of moving fan blades called rotors and fixed blades called stators. The compressor draws air into the combustion chamber where fuel is sprayed in through the injectors. The fuel-air mixture explodes – the hot exhaust gases rush out of the exhaust at high speed. It is the reaction to the force pushing the gases out which drives the jet engine forward.

Some of the power of the exhaust gases is used to drive a fan called a *turbine* – it is this turbine which drives the rotors of the compressor. When the engine is being started the turbine shaft has to be turned with an electric motor.

35

The gas turbine

The gas turbine, which is shown in the photo and diagram below, is very similar to the jet engine. The power it develops is used to turn a shaft, instead of simply providing thrust:

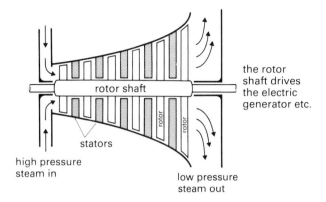

the rotor shaft drives the electric generator etc.

high pressure steam in

low pressure steam out

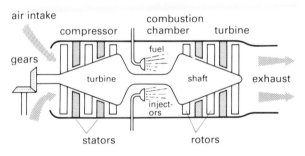

The turbine, which is driven by the exhaust gases, is much bigger than that in the jet engine. All available power from the hot gases is used to drive the turbine. In different situations, this rotary power is used to drive helicopter blades, ships' propellor blades and electrical generators.

The steam turbine

This type of engine is used to produce electricity and to drive large ships. The diagram at the top of the next column shows a simplified cross section of a steam turbine.

High pressure steam is fed into the turbine. As the steam passes through it expands. In expanding, the steam pushes against the rotor blades, and makes the rotor revolve. The photograph shows a steam turbine which will be used to generate electricity.

Exercises

1. Explain with the aid of a diagram how a rocket engine works.
2. Draw a diagram of a jet engine and label its three main parts. Explain how the engine works.
3. What are the main differences between the rocket engine and the jet engine?
4. Draw a labelled diagram of a gas turbine engine and explain how it works.
5. What are the main differences between the gas turbine and the jet engine?
6. Draw a diagram of a steam turbine and explain how it works.

Petrol engines

Petrol engines are used to power cars, motor bikes, vans, and small lorries. Petrol is mixed with air and exploded inside the engine in a *cylinder*. The explosion is used to force down a closely fitting *piston*. The *crankshaft* turns the up and down movement of the piston into rotary movement – this is used to drive the vehicle. An up-movement or a down-movement of the piston is called a *stroke*:

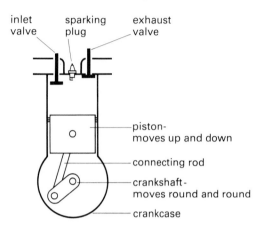

inlet valve sparking plug exhaust valve

piston- moves up and down

connecting rod

crankshaft- moves round and round

crankcase

The four stroke petrol engine

This is by far the most common type of petrol engine. It is called a four stroke engine because there is only one power stroke in every four strokes. The other strokes suck petrol and air in, compress the mixture, and push the burnt gases out. The sequence of the four strokes is:

1. Inlet. The inlet valve opens and the piston moves down. Petrol and air are drawn into the cylinder.

2. Compression. Both valves shut. The piston moves up compressing the petrol and air.

3. Explosion. Both valves are still shut. As the piston reaches the top of the compression stroke the spark plug sparks and ignites the petrol-air mixture. The explosion produced forces the piston down. This is the power stroke.

4. Exhaust. The exhaust valve opens. The piston moves up and pushes the burnt gases out through the exhaust valve:

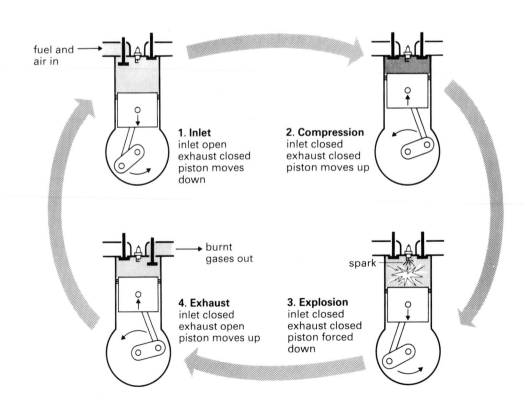

fuel and air in

1. Inlet
inlet open
exhaust closed
piston moves down

2. Compression
inlet closed
exhaust closed
piston moves up

burnt gases out

4. Exhaust
inlet closed
exhaust open
piston moves up

spark

3. Explosion
inlet closed
exhaust closed
piston forced down

36

The two stroke petrol engine

This engine has a power stroke every down stroke. It is much simpler than the four stroke engine as it has no valves. It relies on the piston uncovering holes in the cylinder walls to let petrol and air in, and the exhaust gases out. These holes are called *ports*. The two stroke petrol engine is used mainly for small motor cycles, lawn mowers and portable electric generators. The diagram on the right shows the components of the engine. Note the position of the three openings called the inlet port, the transfer port and the exhaust port.

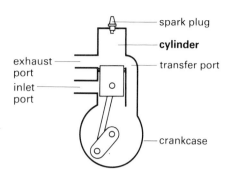

1. Compression and explosion. In the left hand diagram, the cylinder is full of petrol-air mixture, with the piston compressing the mixture. As the piston moves up the cylinder it covers the exhaust and transfer ports. At the same time as this is happening, the inlet port is uncovered and the crank case takes in more petrol-air mixture. When the piston reaches the top the spark plug sparks and explodes the fuel, forcing the piston down.

2. Inlet and exhaust. After the piston has been forced down by exploding fuel, it uncovers the exhaust and transfer ports. The motion of the piston moving down into the crank case forces petrol and air from the crank case, through the transfer port and into the cylinder. The new mixture coming into the cylinder blows the burnt gases out through the exhaust port.

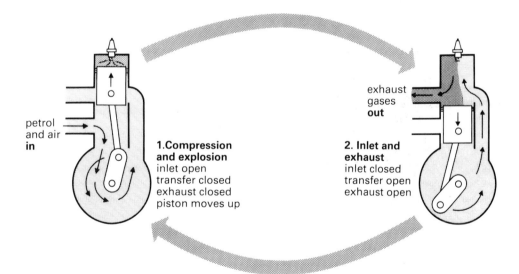

Both the four stroke and two stroke engines have a heavy wheel called a flywheel, which keeps the pistons moving on the unpowered strokes. The two stroke engine is cheaper to manufacture than the four stroke engine because it has fewer moving parts, but it is less efficient than the four stroke engine.

Exercises

1. Draw four diagrams to describe the operation of the four stroke petrol engine.
2. With the aid of diagrams explain how the two stroke petrol engine works.

Diesel engines

Diesel engines are very efficient at pulling heavy loads – they are often used in trucks, buses, trains and ships.

The main difference between the diesel engine and the petrol engine is in the way the fuel is ignited. A diesel engine has no spark plug. When a gas is compressed its temperature rises; in the diesel engine the air is compressed so much that it becomes hot enough to ignite the diesel fuel. The diesel engine has to be made much stronger than the petrol engine to withstand the extra compression needed.

The four stroke diesel engine

This is the most common type of diesel engine. The sequence of the four strokes is:

1. Intake. The inlet valve opens as the piston moves down. Air only is drawn into the cylinder.

2. Compression. Both valves are shut. The piston moves up compressing the air. As the piston nears the top, diesel fuel is sprayed in through the injector.

3. Explosion. Both valves are still shut. As the piston reaches the top of the cylinder the heat of compression is sufficient to ignite the fuel air mixture. The explosion produced forces the piston down – this is the power stroke.

4. Exhaust. The exhaust valve opens. The piston moves up and pushes the burnt gases out through the exhaust valve. Once the piston is in this position, the first stroke is ready to start again.

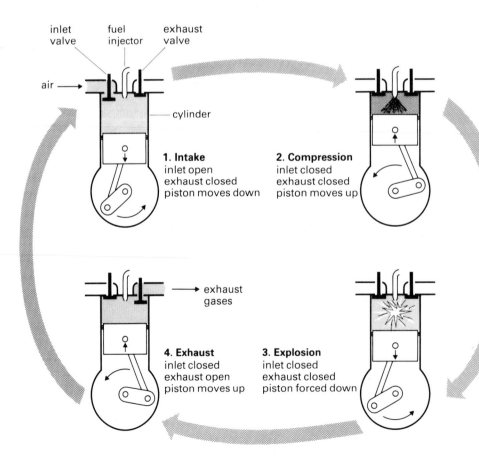

37

The two stroke diesel engine

This engine can be made very large and powerful. It is used mainly to power ships. The two stroke diesel cycle is shown below.

1. Compression and explosion. Starting with the piston moving up: The piston compresses the air in the cylinder, and makes it hotter. Just before the piston reaches the top, diesel fuel is sprayed in through the injector. The heat of compression ignites the fuel and the explosion forces the piston down.

2. Intake and exhaust. As the piston is forced down by the exploding diesel fuel, it uncovers the exhaust and inlet ports. Compressed air is blown in through the inlet port. The compressed air blows the burnt gas out through the exhaust port, and fills the cylinder with fresh air. The piston then starts to move up again and the cycle is repeated.

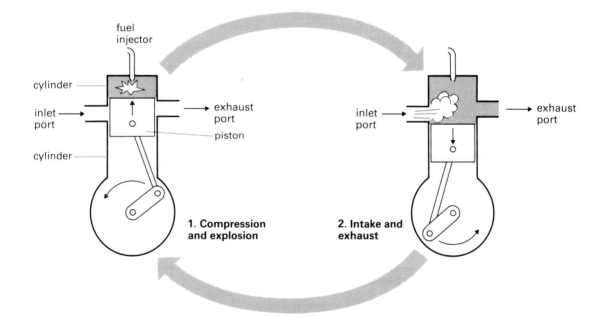

4-stroke diesel engine

The two stroke diesel engine is much simpler than the four stroke as it has fewer moving parts. The two stroke diesel does, however, need a compressor to blow air into the cylinder.

Exercises

1. Draw four diagrams to describe the operation of the four stroke diesel engine.
2. With the aid of diagrams explain how the two stroke diesel engine works.

Questions on chapter 5

1. Draw sketches to show how the effects of thermal expansion and contraction are allowed for in each of the following.
  i Long lengths of pipe
  ii Railway lines.
  iii Electric power cables. (E.A.E.B.)

2. a A bimetallic strip is two pieces of different metal bonded (stuck) together. Brass and iron are bonded together as shown in the diagram.

A iron
brass

  i Draw and label a diagram to show what happens when the strip is heated.
  ii Explain briefly why the change you have shown takes place.
  b One of the major uses of the bimetalic strip is in a thermostat. The diagram below shows a thermostat in an electric iron.

B

  i What are the parts labelled A, B, and C?
  ii Describe how the thermostat will operate so that man-made fibres can be ironed at low temperatures and linen at high temperatures.
  c Give three other uses of the bimetal strip. (A.L.S.E.B.)

3. Liquids expand when they are heated.
  i Sketch a graph of *density* (y-axis) against temperature (x-axis) from 0 °C to 15 °C for a liquid such as alcohol.
  ii On the same axes sketch the graph for water, marking any temperature which you think is significant.
  iii Explain, by reference to your graph, the movement of some water in a beaker as it cools from 15 °C down to freezing point. (W.M.E.B.)

4. a With the aid of a labelled diagram explain how the upper fixed point may be determined for a mercury-in-glass thermometer.
  b State three advantages of using mercury in thermometers.
  c In an ungraduated thermometer, the length of the mercury thread was 2·5 cm when the thermometer was immersed in ice at 0 °C and 25 cm in steam at 100 °C.
    i Using graph paper, plot a graph of temperature against length of mercury thread, assuming that the expansion is uniform.
    ii What temperature corresponds to a thread length of 8·5 cm?
    iii What change in thread length was produced when the temperature changed from 40 °C to 80 °C? (W.J.E.C.)

5. The diagram shows a section through a metal radiator filled with hot water. The radiator is connected to a wall.

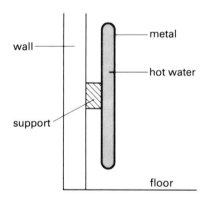

  i How does the heat travel from the water to the air?
  ii Explain, in detail, the process by which the warmed air heats the room.
  iii How does a person sitting directly in front of the radiator get warmed? (E.M.R.E.B.)

6. Give a brief scientific explanation for:
  i Exposed cold water pipes bursting in very cold weather.
  ii String vests keeping you warm even though they contain a large number of holes. (W.J.E.C.)

7. The diagram shows a vessel which is used to prevent hot liquids from going cold. It can also be used to keep cold liquids cool in hot weather.

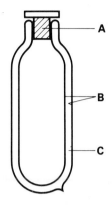

a What is the name of this vessel?
b i What is the part labelled **A** usually made from?
ii Why is this material chosen for this part?
c The walls of this vessel are made of thin glass but its inner surfaces, **B**, are specially treated.
i What is special about these surfaces?
ii Why are these surfaces treated in this way?
d i What is contained in the space **C**?
ii What is the purpose of the space **C**?
(N.W.R.E.B.)

8. The diagram shows a turbo jet engine.

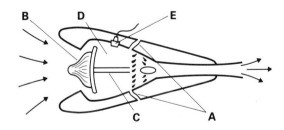

i Write down the letters **A** to **E** and beside each put the name of the part to which the letter refers.
ii Describe the working of this engine.
iii What is the purpose of the turbine in this engine?
iv Explain why the engine is suitable for travel at very high altitudes and yet not suitable for travel in space. (M.R.E.B.)

9. Jet engines can only be used in the Earth's atmosphere whilst rocket engines can also be used in space. Explain the difference between the two engines which makes this possible. (E.A.E.B.)

10. Copy the following table and complete it to show whether the inlet and exhaust valves are open or closed during each stroke of the cycle in a four-stroke petrol engine.

Stroke	inlet valve	exhaust valve
Intake		
Compression		
Power		
Exhaust		

(S.W.E.B.)

11.

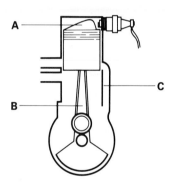

The diagram shows the cylinder of a two-stroke engine. The piston is at its highest position.
i Name the parts marked **A**, **B**, and **C**.
ii Which stroke is just commencing?
iii What is the purpose of part **A**?
iv Why is the fuel/air mixture made to pass through the crankcase?
v What is the purpose of the flywheel?
vi Why does a single cylinder two-stroke engine run more smoothly than a single cylinder four-stroke? (W.M.E.B.)

12. Explain the similarities, and the differences, between 4-stroke petrol and 4-stroke diesel engines.

Light travels in straight lines

The spotlight beams in the photograph suggest that light travels in straight lines.

It is only possible to see objects clearly *because* light travels out from them in straight lines. Every point on an object emits thousands of light rays in all directions. The top of the arrow in the diagram is in one such point, and only a fraction of the rays are marked:

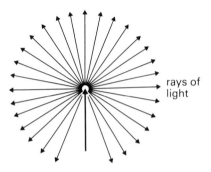

rays of light

The eye picks up one small part of all the possible rays in order to see the tip of the arrow. If the eye moved, the selected part of the rays would change:

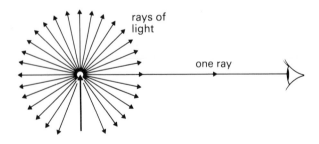

rays of light

one ray

To see the whole arrow, the eye picks up one ray from each point on the arrow:

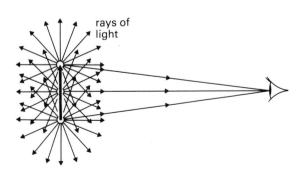

rays of light

The pinhole camera

This is the simplest type of camera – even this would not work if light did not travel in straight lines. It consists of a box with a pinhole in one side, and a screen on the opposite side. The pinhole side is pointed at the object to be viewed. The pinhole selects one ray of light from each point on an object, to form a likeness, or *image* of it on the screen. The diagram below shows how the pinhole selects one ray from the top of the candle:

rays of light

screen

pin hole

image

38

In order to simplify the diagram *only the rays selected by the pinhole are drawn.* The paths of three such rays are drawn in the diagram below showing how an inverted (upside down) image of the object is formed.

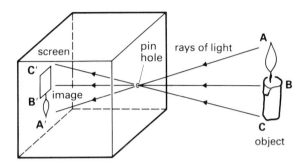

Size of the image

The size of the image depends on two things – the distance of the object from the pinhole, and the distance of the pinhole from the screen.

1. The image on the screen is larger when the object is closer to the pinhole:

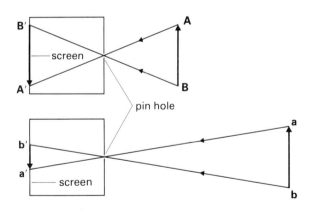

2. The image is larger when the screen is further from the pinhole:

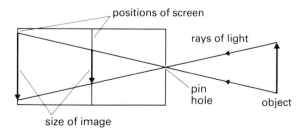

Effect of two pinholes

If there are two pinholes in the camera each pinhole selects its own rays of light from the object making two images on the screen:

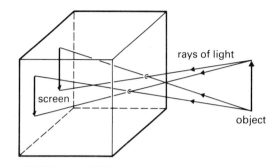

Effect of the size of the pinhole

If the size of the pinhole is increased, a cone of rays is let through instead of a single ray. Because more light is let through, the image is brighter. But each point on the object appears as a small circle of light on the image, so the image becomes blurred.

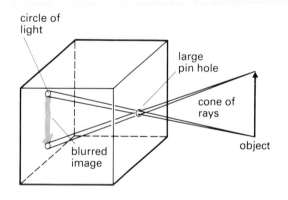

Exercises

1. Draw a diagram of the pinhole camera. Label the screen and the pinhole. Using an arrow to represent the object, draw four rays of light from it, showing how the image forms.

2. Complete this passage using the words below it:
The image in the pinhole camera is *. The * is larger when the object is * the pinhole. If the pinhole is made * the image is *. If two pinholes are made, * images form.
(blurred, image, nearer to, two, inverted, larger).

3. Explain with the aid of a diagram how a larger pinhole affects the image in the pinhole camera.

Shadows and eclipses

A shadow is formed on a surface when an object stops light from reaching it. The type of shadow depends on the size and position of the light source and the object.

Shadow from a point source of light.

A pinhole is used to produce a point source of light:

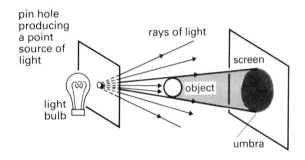

Rays of light are emitted from the point source in all directions. The object prevents all rays which lie between the two marked as thick ones from reaching the screen. The shadow is completely dark, and its edge is sharp. This type of shadow is called an *umbra*.

Shadows from large sources of light

When a large light source is used, the shadow is neither completely dark, nor completely sharp. The type of shadow depends on the relative size of the object and the source:

Source smaller than the object. The shadow on the screen has a dark centre surrounded by a fuzzy half lit area:

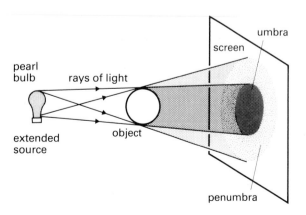

The dark centre is again called the *umbra,* and the part-lit area is called the *penumbra*.

The object cuts *all* light off from the umbra, and only cuts off *part* of the light from the penumbra. Looking from the bottom of the penumbra, for example, the bottom of the bulb would be visible but the top of it would be obscured by the object.

Source larger than the object. The type of shadow in this case depends on the distance of the screen from the object. The umbra forms a small cone of darkness behind the object. If the screen cuts through the umbra, as in diagram **1**, the shadow has both an umbra and a penumbra:

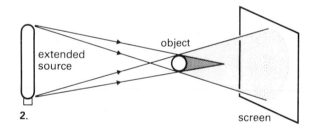

1.

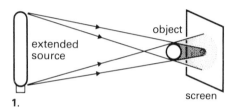

2.

However, if the screen is placed beyond the region of the umbra, as in diagram **2**, the shadow has only a penumbra. Some part of the source is visible from all parts of the penumbra.

Eclipse of the sun

The sun is about four hundred times larger than the moon, but the sun is also about four hundred times further away from us. For this reason the sun and the moon appear the same size in the sky. When the moon passes in front of the sun, it blocks out the light coming from the surface of the sun. The halo of pale white light around the eclipsed sun in the photograph at the top of the next page is due to the sun's atmosphere, and is called the corona:

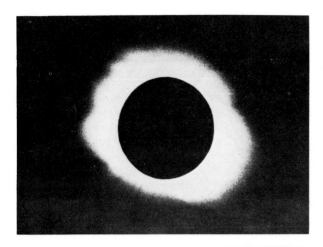

39

The diagram below shows how an eclipse of the sun occurs. A person standing in the umbra can see no part of the sun. This is called a *total eclipse*. A person standing in the penumbra can see a part of the sun. This is called a *partial eclipse*:

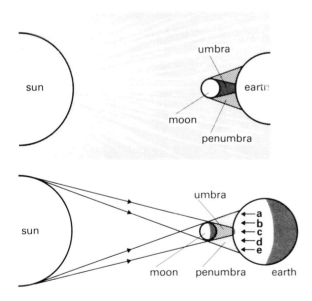

The appearance of the sun to people standing at the positions **a, b, c, d** and **e** on the earth's surface are shown below:

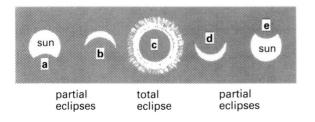

partial
eclipses

total
eclipse

partial
eclipses

Eclipse of the Moon

An eclipse of the moon occurs when the moon passes through the shadow of the earth:

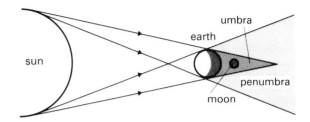

The moon should be completely invisible when it is in the umbra of the earth as no light from the sun should reach it. However, the moon *is* visible – it appears a coppery colour. This is because the atmosphere of the earth acts like a lens, and bends some of the sun's light so that some of it reaches the moon:

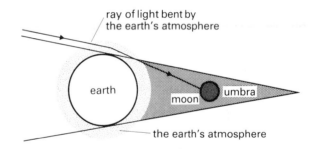

Exercises
1. Draw a diagram to show how the shadow of an object is formed on a screen by a small, extended source. Explain what is meant by the terms umbra and penumbra, and mark them on your diagram.
2. Draw a diagram to show how an eclipse of the sun occurs. Mark the regions on the earth, where the eclipse appears total and partial.
3. Draw a diagram to show how an eclipse of the moon is formed. Explain why the moon is still visible when it is in the umbra of the earth, and no light from the sun should reach it.

Reflection of light from plane mirrors

The way in which light is reflected from a surface depends on the nature of the surface.

A mirror has a very smooth surface and the rays of light falling on it are reflected straight off it. This is called *regular reflection:*

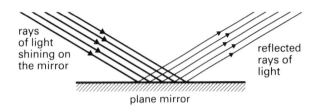

rays of light shining on the mirror

reflected rays of light

plane mirror

Light is also reflected from the paper of this book. The surface of the paper is much rougher than a mirror's. The light is reflected off in all directions, depending on the angle of the part of the surface it hits. This is called *diffuse reflection.* Note that in order to make the diagrams more clear, in this book rays coming to a surface are drawn thicker than the rays leaving it:

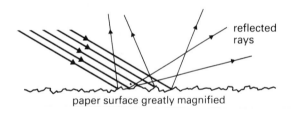

reflected rays

paper surface greatly magnified

Reflection: names
There are a number of words with special meanings that are used in studying reflection:

Incident ray. This is the ray of light that shines onto the mirror.

Normal. This is the name given to a line drawn at 90° to the mirror, drawn from the point where the incident ray strikes the mirror.

Angle of incidence. This is the angle between the normal and the incident ray.

Reflected ray. This is the ray of light which is reflected off the mirror.

Angle of reflection. This is the angle between the normal and the reflected ray.

These are all marked on the next diagram.

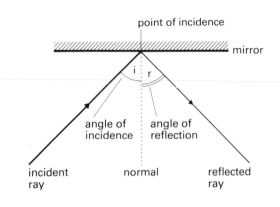

point of incidence

mirror

angle of incidence

angle of reflection

incident ray

normal

reflected ray

Laws of reflection
There are two laws of reflection.

Law 1. If the incident ray, the normal and the reflected ray were marked by metal rods, then it would be seen that a flat piece of paper could be laid across them. In other words:

The incident ray, the reflected ray and the normal at the point of incidence all lie in the same plane:

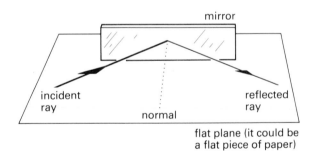

mirror

incident ray

normal

reflected ray

flat plane (it could be a flat piece of paper)

Law 2. This law may be demonstrated by using a ray box to make rays of light hit a mirror at different angles. When the angles of incidence and angles of reflection are measured they are found to be equal.

The angle of reflection is equal to the angle of incidence:

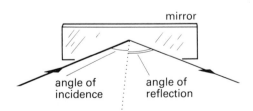

mirror

angle of incidence

angle of reflection

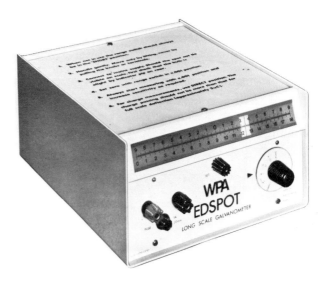

40

Both regular and diffuse reflection, in a natural setting.

Using rotating mirrors

If a ray of light is directed at a mirror as shown in diagram **1**, and the mirror is then rotated as in diagram **2**, it is found that the direction of the reflected beam changes. The reflected beam turns through twice the angle that the mirror is turned:

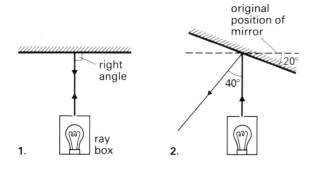

This idea is used in some sensitive electric meters such as the Edspot galvanometer shown in the photograph at the top of the next column.

This meter has a small mirror instead of a pointer. A beam of light is shone at the mirror:

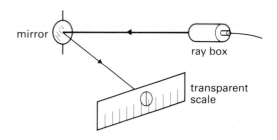

The beam of light is reflected off the mirror onto a scale. Unlike a pointer, a beam of light has no weight, so the scale can be placed a long way from the mirror. A small movement of the mirror makes the end of the light beam move a long way across the scale, so the meter is very sensitive.

Exercises

1. Make a list of the following terms with their meanings: incident ray, reflected ray, normal, angle of incidence, angle of reflection. Learn the list!
2. State the two laws of reflection of light.
3. Describe an experiment to demonstrate the second law of reflection. Draw a diagram of the apparatus marking on it the angles, which are to be measured.
4. Explain the difference between regular and diffuse reflection.
5. If a mirror is turned through 17° what angle is the reflected beam turned through?

The image in plane mirrors

The way in which light reflects from a normal flat mirror, or *plane* mirror, deceives the eye. Rays of light coming from an object in front of the mirror are bent by it and look as though they have come from behind it! The image is not a *real image* that could be picked up on a screen, but a non-real or *virtual image,* at the place where the light *appears* to be coming from:

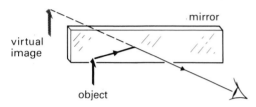

The position of the image
The image in the mirror is as far behind the mirror as the object is in front of it; the image lies directly opposite the object. This may be shown using a small mirror, and two large pins that are taller than the height of the mirror.

One of them is placed in front of the mirror – this is called the *object pin.* The image of the bottom half of it should be seen in the mirror. The second pin is the *finder pin* – this is placed roughly where the image appears to be:

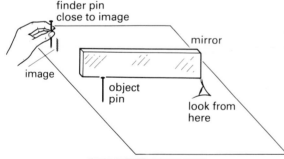

If the finder is not in the right place, the image of the bottom of the object pin will not line up with the top of the finder pin. As it is gradually moved to the right position, the finder and the image will be in line no matter what angle they are viewed from. The finder is then in the same position as the image. When this is done it is found that:

The image is as far behind the mirror as the object is in front;

A line joining the object to the image is at right angles to the mirror:

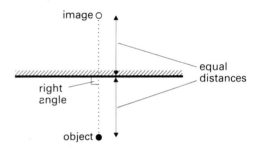

The type of image
This section is only about flat, or plane, mirrors. For a plane mirror:

The image in a plane mirror is virtual.

The image in a plane mirror is the same size as the object.

The image in a plane mirror is laterally inverted.

"Laterally inverted" means that the image is changed round from side to side. The left hand side of the object is seen as the right hand side of the image. The girl in the photograph below is holding a poster in her left hand. But in the image the poster is being held in the right hand:

41

Ray diagram for a plane mirror

A ray diagram can be used to show the actual path taken by light rays. The diagram is drawn in three steps:

1. Draw the mirror, the object and the image in the right places:

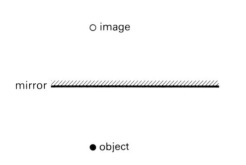

2. Draw in the two rays of light which appear to leave the image and enter each of two eyes. Draw the lines dashed behind the mirror – this is where the light rays appear to come from having been reflected by the mirror:

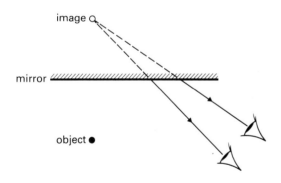

3. Complete the diagram by drawing in the incident rays which went on to form the reflected rays. The paths taken by the rays is now complete:

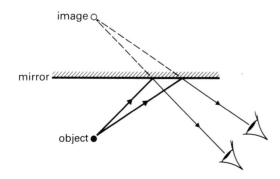

The periscope

This is used for looking over the top of obstacles. It is made from two plane mirrors at an angle of 45°. The diagram shows two parallel rays of light travelling through it:

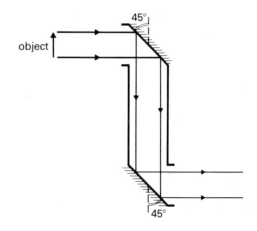

Exercises

1. Copy out the following sentences filling in the missing words from the lists.

The image in a * mirror is as far * the mirror as the * is in *, and a * joining the * to the object is at * angles to the mirror.
(front, right, image, plane, object, behind, line).

The * in a plane mirror is the same size as the *, *, and laterally *.
(virtual, image, inverted, object).

2. What is meant by "laterally inverted"?
3. The diagram shows an object in front of a plane mirror. Copy it, and draw the image.

4. An object is placed 4 cm in front of a plane mirror. Draw to scale a ray diagram to show how the image is formed.
5. Draw a diagram of a periscope, showing *three* parallel rays of light through it.

Refraction

Light rays bend as they pass through from one substance (such as air) to another substance (such as glass). This bending is called *refraction* and an example of it is shown in the illustration above!

The letters appear to have moved because the rays of light from them bend as they come out of the glass block. The eye is deceived into thinking that the letters are in a different position.

Refraction: names

There are a number of words with special meanings used in studying refraction. They are the same as those for reflection with two new ones added:

Refracted ray. This is the ray which has been bent at the surface.

Angle of refraction. This is the angle between the refracted ray and the normal:

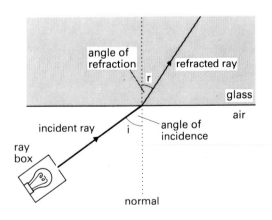

Refractive index. The refractive index is a measure of the ability of a substance to bend light. The refractive index is different for different substances: e.g. the refractive index of glass is 1.5 and of water is 1.3.

Laws of refraction

There are two laws of refraction:

Law 1. If the normal, and the routes of the rays of light were marked by rods of metal then a flat piece of paper could be laid across them. In other words:

The incident ray, the refracted ray, and the normal at the point of incidence all lie in the same plane:

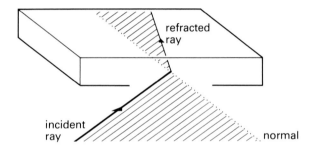

Law 2. The amount of bending that the beam undergoes depends on the substances between which the beam is travelling. No matter what the angle of incidence is, it is found that:

The sine of the angle of incidence divided by the sine of the angle of refraction stays constant. This constant is the refractive index.

Stated mathematically:

$$\text{refractive index} = \frac{\text{sine of angle of incidence}}{\text{sine of angle of refraction}}$$

$$n = \frac{\sin i}{\sin r}$$

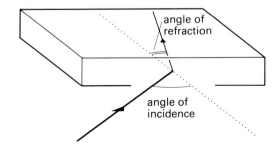

42

Demonstrating the second law

The second law may be demonstrated by using a ray box to make rays of light hit a semi-circular glass block at different angles, as shown in the diagram:

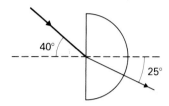

The angles of incidence and refraction are measured. When the sine of the angle of incidence is divided by the sine of the angle of refraction, this always gives the same number. The number is the refractive index of the substance. In general, the refractive index increases with the density of the substance.

Real and apparent depth

Swimming pools always appear shallower than they really are: The same effect happens with a glass block – rays of light from the bottom of the block are bent outwards as they pass from the glass to the air as shown:

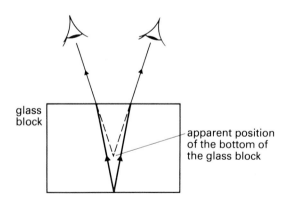

glass block

apparent position of the bottom of the glass block

The bending of the light rays deceives the eye into thinking that they are coming from a point nearer the surface than they really are.

Since the change in depth depends on the refractive index of the substance, this can be used to measure its refractive index. It is worked out from the formula:

$$\text{refractive index} = \frac{\text{real depth}}{\text{apparent depth}}$$

Refractive index by the real and apparent depth method.

An object pin is placed under a glass block, and viewed down through the block. It looks nearer through the glass. A finder pin is placed against the side of the block, and is moved around until the finder and the image line up with one another no matter what angle they are viewed from. The finder is then marking the position of the image. The real and apparent depths of the glass block are measured with a rule and the refractive index is found by substituting in the formula given at the bottom of the last column.

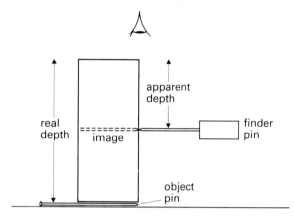

apparent depth

real depth

image

finder pin

object pin

Exercises

1. Write down the meanings of the terms *refracted ray* and *angle of refraction* – then learn them too!

2. State the two laws of refraction of light.

3. Describe with the aid of a diagram how you would demonstrate the second law of refraction.

4. Work out the refractive index for the material in this semicircular block:

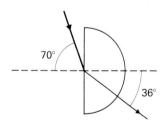

5. You are given a beaker of water with a pin in the bottom, a rule, and another pin that will slide up and down a stand. Explain with the aid of a diagram how you would find the refractive index of water.

6. A glass block is 6 cm long but when you look through the glass it seems to be 4 cm long. What is the refractive index of the glass?

Refraction and reflection in transparent blocks

The way light passes through glass blocks depends on the shape of the block, and the angle at which the light enters the block. The light may pass through the block, in which case it will be *refracted* at both surfaces – or it may be totally *reflected* from the second surface.

Refraction in glass blocks

When a ray of light passes from a less dense substance to a more dense one, (i.e. on entering a glass block) the beam is bent *towards* the normal. As it leaves the block, it is bent *away* from the normal. The actual path taken by the ray will depend on the shape of the block.

Rectangular glass block. A ray of light meeting a block of this shape is refracted once as it enters the block, and a second time as it leaves it. The ray that comes out is called the *emergent ray*. The emergent ray is parallel to the incident ray, as in diagram **1**:

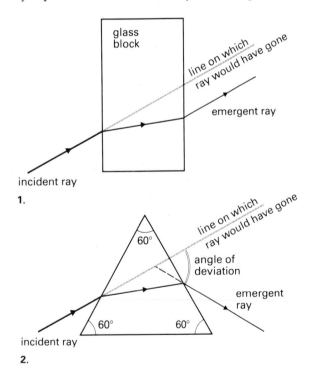

1.

2.

Triangular glass block. The ray of light in diagram **2** is bent at each surface, but the ray is no longer parallel because the surfaces are not parallel to one another. The path of the ray of light has been changed, or *deviated*. The angle between the emergent ray and the incident ray is called the angle of deviation.

Total internal reflection

If a ray of light is shone at the centre of a semicircular block as shown, most of the light is refracted at the straight face, and passes on through. But there is a weak ray of light that reflects from the surface:

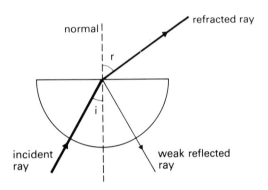

As the angle of incidence is increased two things happen. The angle of refraction increases, but the strength of the refracted ray gets smaller. At the same time, the reflected ray becomes stronger. Eventually a point is reached when the angle of refraction is 90°. The corresponding angle of incidence is called the *critical angle*:

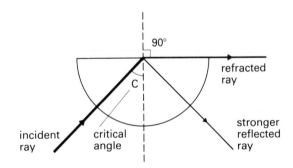

If the angle of incidence is increased further, no light passes through the surface and the ray of light is *totally internally reflected*:

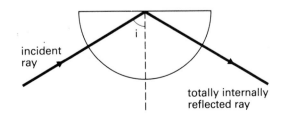

43

Total internal reflection only occurs when the ray of light is going from a *more dense* medium such as perspex or glass, to a *less dense* medium such as air.

A material such as diamond, shown in the photograph on the right, has a high refractive index. This means that the critical angle is small, so that light is easily reflected. Much of the light entering a carefully cut diamond is totally internally reflected. It is this reflection which makes a diamond sparkle in the light.

Reflecting prisms
Right angled prisms can be used to reflect beams of light. They can deflect beams through 90° and 180° as shown:

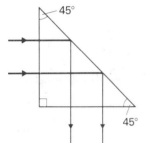

turning rays of
light through 90°

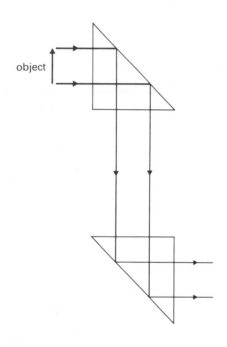

object

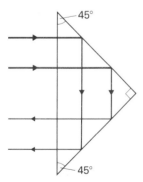

turning rays of
light through 180°

Light is much more efficiently reflected by total internal reflection than by a mirror. For this reason prisms are used in high quality periscopes like the ones used in submarines. The diagram in the next column shows the paths of two parallel rays of light as they travel through a prismatic periscope.

Exercises
1. Draw diagrams to show the passage of a ray of light through a rectangular glass block and a 60° triangular glass block.
2. Explain the meaning of the terms: "critical angle" and "total internal reflection". Use diagrams to help your explanation.
3. Draw a diagram of a prismatic periscope, marking the paths of three parallel rays of light through it.

Questions on chapter 6

1. The diagram shows an object in front of a pinhole camera.

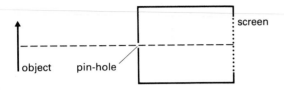

   a Draw a similar diagram in your book and:
    i Draw one ray from each end of the object to the screen.
    ii Draw the image on the screen.
   b How can the camera be slightly modified so that the image produced will be larger?
   c What would be the effect on the image of making two pinholes in the camera?   (N.W.R.E.B.)

2. A lamp is used to illuminate an object, a ball, in order to cast a shadow on a screen as shown in the diagram:

   a What are the names of the two shadow areas marked **C** and **D**?
   b Draw a similar diagram in your book, drawing on it two rays from point **A** and two rays from point **B** which just touch the ball and pass onto the screen, showing how the two shadow areas are formed.
   c If the arrangement in the diagram were regarded as an eclipse of the sun, involving the Sun, Earth, and Moon, which object would represent:
    i The Sun
    ii The Earth
    iii The Moon.   (M.R.E.B.)

3. a Draw a diagram which illustrates an eclipse of the sun. Clearly mark and label.
    i A position where a *total* eclipse might be observed.

    ii A position where a *partial* eclipse might be observed.
   b Draw diagrams which illustrate the differences between the reflection of light from a matt surface and a polished surface. Label the surfaces.
   c State the laws of reflection.
   d Draw a diagram of a simple periscope using mirrors, showing two parallel rays from a distant object through the system. State whether the final image is the same way up as the object or upside down.   (E.A.E.B.)

4. i The diagram shows parallel rays of light striking a plane mirror.

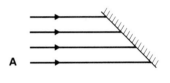

Copy the diagram and draw in the reflected rays.
ii Copy the diagram below and sketch in the image of the object formed by the plane mirror.   (M.R.E.B.)

5. The diagram below shows an object placed in front of a plane mirror. Copy it into your book.

i Mark accurately the position of the image.
ii Draw two rays from the object, striking the mirror, and then entering the eye.
iii What is meant by lateral inversion?   (W.M.E.B.)

6. Why does light change direction when it goes from one material to another?
   Copy the diagrams showing rays of light entering two semi-circular glass blocks:

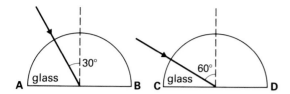

Mark on your diagrams the paths of the various light rays after the incident rays strike the boundaries **AB** and **CD**. Do not forget to add appropriate arrows to your rays. (You are not expected to calculate angles of refraction). (W.M.E.B.)

7. In an experiment to measure the refractive index of a glass block a boy set up the following apparatus.

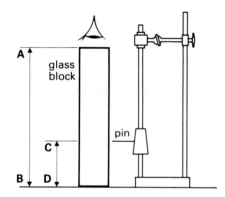

i Describe what the boy did in the experiment.
ii What measurements did he take?
iii If the results of his experiment were AB 12 cm, CD 4 cm calculate the refractive index of glass. (A.L.S.E.B.)

8.

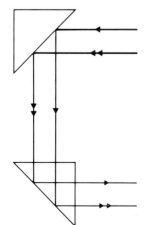

The last diagram shows two glass prisms being used *incorrectly* to make a simple periscope.

a i Re-draw the prisms correctly placed relative to each other to produce an erect image.
ii Draw on your diagram two rays of light passing completely through the instrument.
iii Mark on your diagram any angles which are important in the reflection or refraction of the rays of light.
b Why are prisms preferred to mirrors when making a good quality periscope? (N.W.R.E.B.)

9.

Three rays of light are spreading from point **E** on the bottom of a tank of liquid as shown above.
i What name is given to the bending of the ray at **A**?
ii What name is given to the dotted line from **H** to **B**?
iii What is happening to the ray at **C**?
iv What name is given to the acute angle **D**?
v Measure angle **D** using a protractor and write down the result.
vi Explain why the ray behaves as it does at point **C**.
vii Looking into the liquid from **K**, where would the bottom of the tank appear to be?
viii Use a ruler to make the appropriate measurements and then calculate the refractive index of the liquid.
ix In what direction must a ray leave **E** so that it would not be bent at the surface?
x Looking from **J** towards **C**, what would be the appearance of the liquid's surface? (Y.R.E.B.)

Lenses and curved mirrors

This section introduces some of the words used in the study of curved mirrors and lenses.

Types of rays of light
Rays of light may be one of three types: *converging rays* get closer together; *parallel rays* stay the same distance apart; *diverging rays* get further apart:

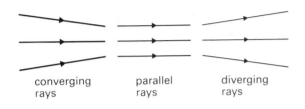

converging rays   parallel rays   diverging rays

Types of lens
A lens is made of a piece of glass which is curved on both sides. The shape of the surfaces could be made by cutting circular pieces out of a plastic football. There are two types of lens: A *convex lens* bulges at the centre; a *concave lens* gets thinner at the centre:

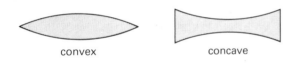

convex   concave

The optical centre of the lens is the centre of the glass of the lens.

The principal axis of a lens is a line drawn through the optical centre at right angles to the lens.

A convex lens causes rays of light to converge. A concave lens causes light to diverge:

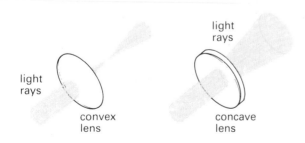

light rays   convex lens   light rays   concave lens

For simplicity all further rays diagrams will be drawn in cross-section.

Principal focus and focal length of a lens
A place where rays meet or where rays appear to come from is called a *focus*. Both types of lens produce focuses.

Principal focus of a convex lens. This is the point where rays of light parallel to the principal axis meet after being bent at the lens:

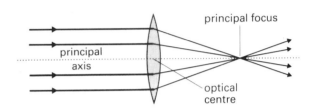

principal focus

principal axis

optical centre

Principal focus of a concave lens. This is the point from which rays of light parallel to the principal axis appear to have come, after being bent by the lens:

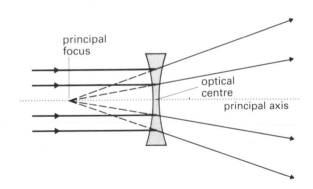

principal focus

optical centre

principal axis

Focal length of a lens. This is the distance between the principal focus and the optical centre of the lens. The focal length of both the convex and concave lenses can be seen in the diagrams above. The shorter the focal length of the lens, the "stronger" it is – its ability to bend light rays is greater.

Types of mirror
There are two types of mirror: a concave mirror and a convex mirror. The shapes of their surfaces may be obtained by cutting circular pieces out of a plastic

football. Reflection takes place at the inside surface in a concave mirror and at the outside surface in a convex mirror:

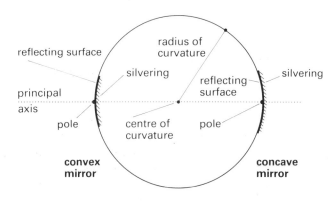

Centre of curvature. This is the centre of the sphere of which the mirror forms a part.

Radius of curvature. This is the distance from the centre of curvature to the mirror.

Pole of a mirror. This is the centre of the piece of glass which makes the mirror.

Principal axis. This is a line joining the pole of the mirror to the centre of curvature.

Concave mirrors cause rays to converge and convex mirrors cause rays to diverge:

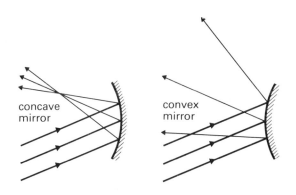

Principal focus and focal length of a mirror.
As with lenses rays of light may be brought to a focus:

Principal focus of a concave mirror. This is the point where rays of light parallel to the principal axis meet after reflection at the mirror:

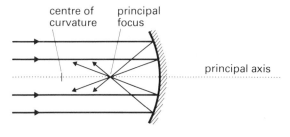

Principal focus of a convex mirror. This is the point from which rays of light parallel to the principal axis *appear* to have come:

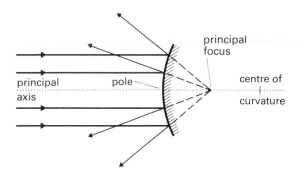

Focal length of mirrors. This is the distance between the principal focus and the pole of the mirror. From the diagrams above it can be seen that:

The radius of curvature is twice the focal length.

Exercises
1. Draw a diagram to show how two rays of light parallel to the principal axis are bent by (a) a convex lens, and (b) a concave lens. Mark on each diagram the principal focus and the optical centre.
2. Draw a diagram showing how two rays of light parallel to the principal axis are reflected from (a) a concave mirror and (b) a convex mirror. Mark on each diagram the centre of curvature, principal focus and pole of the mirror.

Images in lenses

Lenses are used in telescopes, microscopes, cameras, spectacles and many other instruments. The image formed by a lens depends on the position of the object and the type of lens. The image can be:

Virtual or real. – on the same side of the lens as the object if it is virtual, but on the opposite side of the lens to the object if it is real.

Magnified or diminished. – either larger or smaller than the object.

Upright or inverted. – either the right way up or upside down.

Concave lens

The image formed by a concave lens is always of the same type no matter where the object is placed. The image is virtual, diminished, upright, and on the same side of the lens as the object:

Convex lens

The nature of the image depends on the position of the object with respect to the principal focus.

Nature and position of the image. This can be investigated using the apparatus shown at the top of the next column.

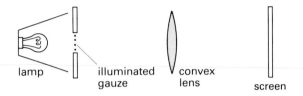

The illuminated gauze is placed at different distances from the lens – the screen is then moved until the image is focussed on it. Real images are only produced when the gauze is further from the lens than the principal focus. When the gauze is a long way off the image is close to the lens and small. As the gauze is moved towards the lens, the image moves further away and becomes bigger. The results of this experiment are summarised in the diagram. **F** is the principal focus and **2F** is a reference point at a distance of twice the focal length:

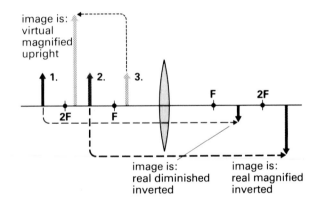

Object beyond 2F. This is the arrow in position **1**. The image is real, diminished, and inverted; it lies between **F** and **2F**.

Object between 2F and F. This is the arrow in position **2**. The image is real, magnified, inverted and lies beyond **2F**.

Object between F and the lens. This is the arrow in position **3**. In this case the image is virtual, erect, magnified, and on the same side of the lens as the object.

The photograph shows the magnified, virtual image formed when the object (the lined paper) is close to the lens. In this case the convex lens is being used as a magnifying glass:

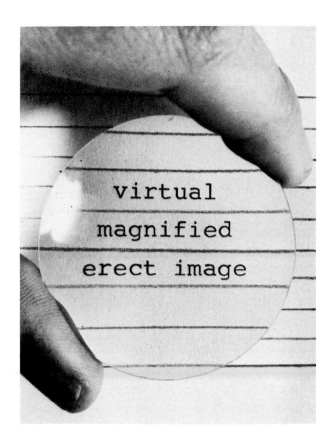

45

The focal length, which is the distance between the paper and the lens, is measured with the rule. If the experiment is carried out with different lenses it is found that the focal length is shorter when the surfaces of the lens are more curved. A lens with a shorter focal length is thought of as being "more powerful" because it bends the light more:

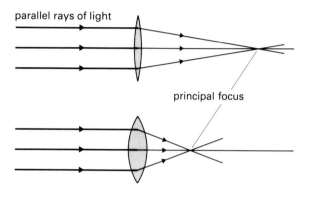

Finding the principal focus. A simple method uses a piece of white paper and a rule. The lens is held in the hand and moved until an image of a window on the opposite side of the room is focussed onto the paper. The parallel light from the distant window is then brought to the principal focus at the paper:

Exercises

1. What type of image is formed by a concave lens?
2. What is the nature (real or virtual, magnified or diminished) of the image formed by a convex lens of focal length 4 cm, when the object is at the following distances from the lens:
(a) 2 cm, (b) 6 cm, (c) 10 cm?
3. Explain how you would measure the focal length of a convex lens.
4. With the aid of diagrams, explain how the power of a convex lens is affected by its shape.
5. What type of lens is being used here? In what position is the object (the stamp)?

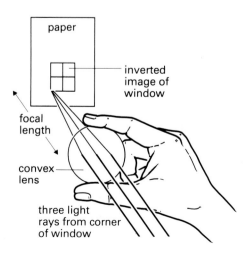

Images in curved mirrors

Curved mirrors are often used in telescopes, projectors and other optical instruments. The image formed by a curved mirror depends on the position of the object and the type of the mirror. Like lenses the image in a curved mirror can be:

Virtual or real – behind the mirror if it is virtual or in front of the mirror it it is real.

Magnified or diminished.

Upright or inverted.

Convex mirrors

Like the concave lens, the image in a convex mirror is always of the same type no matter where the object is. The image is virtual, upright, diminished and behind the mirror.

Because the image in a convex mirror is diminished, the convex mirror gives a wider field of view than a plane mirror. For this reason convex mirrors are sometimes used as driving mirrors in cars. It is possible to see more in a convex mirror, but since the size of the image is smaller, it makes things look further away than they really are. These two effects are shown in these photographs, taken from the drivers seat of a car. The passing cyclist is just visible in the plane mirror, but easily seen in the convex one:

Concave mirrors

Like concave lenses, the nature of the image depends on the position of the object with relation to the principal focus (**F**) and the centre of curvature (**C**).

Nature and position of the image. This can be investigated using the following apparatus:

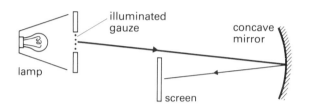

The illuminated gauze is placed at different distances from the mirror the screen is then moved until the image is focussed on it. Real images are only produced when the gauze is further from the mirror than the principal focus. When the gauze is a long way from the mirror the image is close to the mirror and small. As the gauze is moved towards the mirror, the image moves further away from the mirror and becomes larger. The results of the experiment are summarized in the diagram at the top of the next page.

In the plane mirror below, the cyclist's image can be seen in the mirror, but the field of view is restricted. It would be quite easy for the image of the cyclist not to appear at all.

In the concave mirror below, much more of the cyclist can be seen because of the field of view is much wider. Note that the size of the image is smaller.

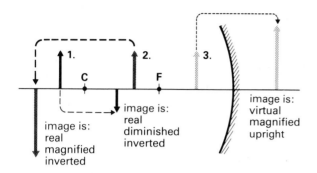

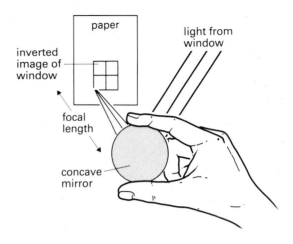

46

Object beyond C. This is the arrow in position **1**. The image is real, diminished, inverted and lies between **F** and **C**.

Object between C and F. This is the arrow in position **2**. The image is real, magnified, inverted, and lies beyond **C**.

Object between F and the mirror. This is the arrow in position **3**. In this case the image is virtual, upright, magnified, and behind the mirror.

The photograph shows the magnified virtual image formed when an object is close to the mirror. A concave mirror is used in this way as a shaving or make-up mirror:

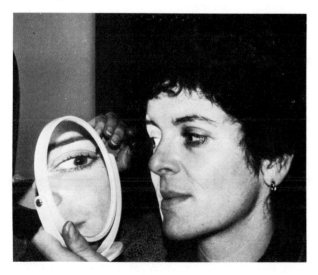

The focal length, which is the distance between the paper and the mirror, is measured with the rule. If the experiment is carried out with different mirrors it is found that the focal length is shorter when the mirror is more curved. A mirror with a shorter focal length is "more powerful" as it bends the light more:

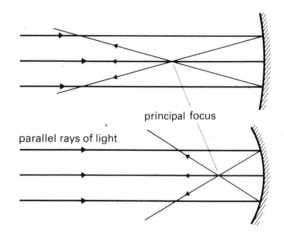

Finding the principal focus. A simple method uses a piece of white paper and a rule. The mirror is held in the hand and moved until an image of a window on the opposite side of the room is focussed onto the paper. The parallel light from the distant window is then being brought to the principal focus at the paper as shown in the next diagram:

Exercises
1. What is a convex mirror used for? What advantage has it over a plane mirror?
2. What is the nature (real or virtual/magnified or diminished) of the image formed by a concave mirror of radius of curvature 8 cm, when the object is at the following distances from the mirror; (a) 2 cm, (b) 6 cm, (c) 10 cm?
3. How would you find the focal length of a concave mirror?
4. Explain using diagrams, how the power of a concave mirror is affected by its shape.

Ray diagrams

The size, nature, and position of the image formed by a lens or mirror can be found by drawing a ray diagram.

Lenses

Two special rays are used to find the image.

1. Centre ray. A ray of light striking the optical centre carries straight on through:

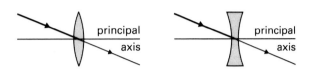

2. Parallel ray. This is a ray of light which is parallel to the principal axis. For a convex lens the parallel ray is bent so that it goes through the far principal focus. For a concave lens the parallel ray is bent so that it *appears to come* from the near principal focus:

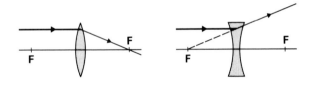

Drawing the diagram. First a 'skeleton' is drawn as shown. The lens is drawn as a vertical line with a small symbol to show the type of lens. The reference point **2F** is twice as far from the lens as the principal focus **F**. The object is drawn as a vertical arrow:

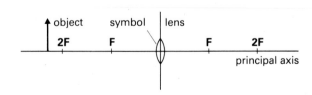

The ray diagram is completed by drawing the *centre ray* and *parallel ray* from the top of the arrow. The top of a real image is formed at the point where the rays cross. If the rays diverge, the point *from which they appear to come* is the top of the *virtual* image.

Ray diagram for a convex lens
Object beyond 2F:

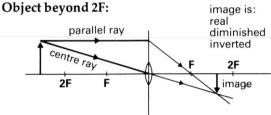

image is:
real
diminished
inverted

Object between 2F and F:

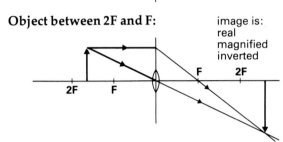

image is:
real
magnified
inverted

Object between F and lens (Magnifying glass):

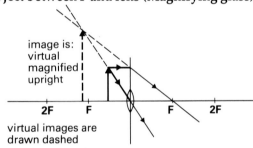

image is:
virtual
magnified
upright

virtual images are drawn dashed

Ray diagram for a concave lens
The only diagram is:

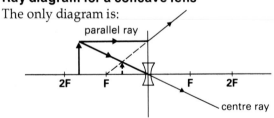

Exercises on lenses
1. A convex lens has a focal length of 3 cm. An object 2 cm high is placed at the following distances from the lens (a) 9 cm, (b) 6 cm, (c) 5 cm, and (d) 1.5 cm. In each case draw a full scale diagram to find the size of the image and its distance from the lens. Is the image real or virtual?

2. A concave lens has a focal length of 4 cm. An object 4 cm high is placed 8 cm from the lens. Draw a full scale diagram to find the size of the image and its distance from the lens.

47

Mirrors

Ray diagram for mirrors are drawn in a similar manner. The same two special rays are used.

1. Centre ray. A ray of light striking the pole of the mirror is reflected back at the same angle:

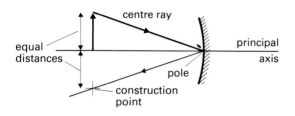

When the object is drawn, a dot is marked under it at the same distance from the principal axis as the top of the object. The path of the centre ray is then drawn by joining the top of the object to the pole, and the pole to the dot.

2. Parallel ray. For a concave mirror the parallel ray is reflected so that it passes through the principal focus. For a convex mirror the parallel ray is reflected so that it appears to *come from* the principal focus:

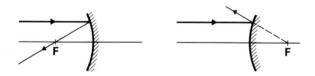

Drawing the diagram. First the skeleton has to be drawn with the mirror, principal focus, and centre of curvature marked on it:

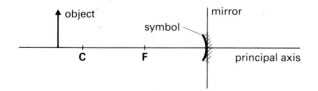

The ray diagram is completed by drawing the centre ray and the parallel ray from the top of the object. The top of a real image is the point where the rays cross, the top of a virtual image is the point where the rays appear to come from.

Ray diagrams for a concave mirror
Object beyond C:

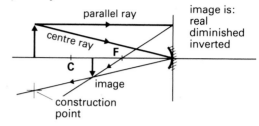

image is:
real
diminished
inverted

Object between C and F:

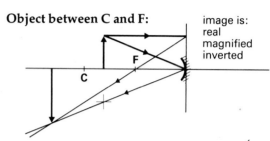

image is:
real
magnified
inverted

Object between F and mirror:
(shaving mirror)

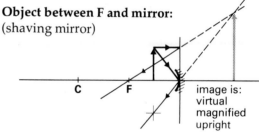

image is:
virtual
magnified
upright

Ray diagram for a convex mirror
The only diagram is:

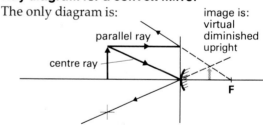

image is:
virtual
diminished
upright

Exercises on mirrors

3. A concave mirror has a radius of curvature of 8 cm. An object 2 cm high is placed at the following distances from the mirror: (a) 12 cm, (b) 8 cm, (c) 6 cm and (d) 2 cm. Draw full scale diagrams to find the size of the image and its distance from the mirror. Is the image real or virtual?

4. A convex mirror has a focal length of 3 cm. An object 4 cm high is placed 5 cm from the mirror. Draw a full scale diagram to find the size of the image and its distance from the mirror.

The eye and its defects

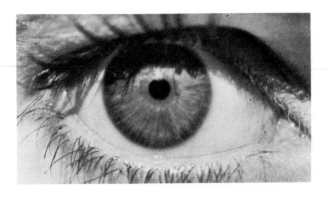

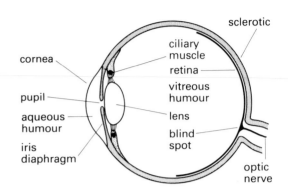

Structure of the eye

The diagram at the top of the next column shows a cross section of the eye. The various parts, and what they do, are explained below.

Sclerotic. The white, leathery outer case of the eye is called the sclerotic.

Cornea. Light enters the eye through the cornea, a curved clear "window".

Aqueous humour. The light then travels through the water-like aqueous humour behind the cornea.

Iris. The coloured part of the eye is called the iris diaphragm.

Pupil. This is the black hole in the centre of the iris through which the light passes. The size of the pupil is controlled by the iris. When it is dark, the iris opens making the pupil large to let more light in. In bright light, the iris closes to make the pupil small.

Lens. There is a lens inside the eye behind the pupil. It is a convex lens and its shape alters to focus the light.

Ciliary muscle. These muscles form a ring round the lens. When they contract they squash the lens into a rounder shape to make it stronger.

Vitreous humour. This is a clear jelly-like substance filling the eyeball.

Retina. The rays of light are focussed onto the retina, a light sensitive layer at the back of the eye. A real inverted image is formed on it.

Optic nerve. The nerve which transmits the image on the retina to the brain is called the optic nerve.

Blind spot. The blind spot is the point where the optic nerve is connected to the retina. Unlike the rest of the retina it is not sensitive to light.

Lenses used in correction of long-and short-sight. Which is which?

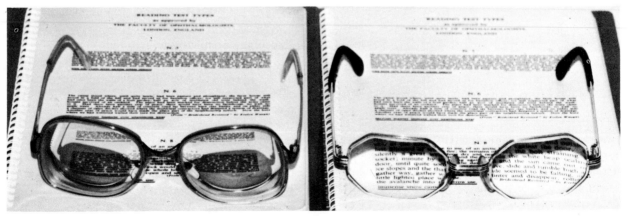

48

1. Normal eye – distant object:

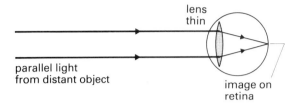

2. Normal eye – near object:

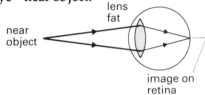

3. Short sight:

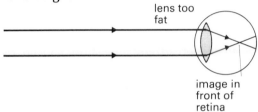

4. Short sight corrected:

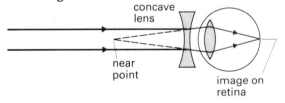

5. Long sight:

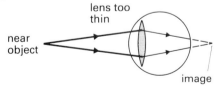

6. Long sight corrected:

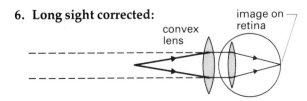

Ciliary muscles and the normal eye

When the ciliary muscles are relaxed, the lens is slim. The relaxed lens has a focal length just sufficient to bring the parallel light from a distant object into focus on the retina – see diagram **1**.

If the object is brought closer to the eye, then the ciliary muscles start to squash the lens to make its surfaces more curved. This gives the lens the shorter focal length that is needed to focus the image onto the retina as shown in diagram **2**.

Short sight

A person with short sight can see near things clearly but distant objects are blurred. His eye is capable of focussing diverging rays from near objects, but his eye lens will not relax enough to focus parallel light from distant objects. The image is formed in front of the retina as shown in diagram **3**.

Correction of short sight. A concave lens is therefore placed in front of the eye to make the light diverge so that it appears to come from a near point. The eye can now cope with this and focuses the rays sharply onto the retina, as in diagram **4**.

Long sight

A person with long sight can see distant objects clearly, but near objects are blurred. His problem is the exact opposite of the short sighted person's. With a close object, the lens will not squash sufficiently to focus the diverging light onto the retina. The image would be formed behind the retina as in diagram **5**.

Correction of long sight. A convex lens is placed in front of the eye to make the light parallel so that it appears to come from a distant object. The eye can now focus this parallel light onto the retina, as in diagram **6**.

Exercises

1. Draw a labelled cross sectional diagram of the eye.
2. Explain how the eye focusses the light from objects at different distances.
3. How is the amount of light entering the eye controlled?
4. Explain what is meant by *short sight* and how it is corrected.
5. What is *long sight*? How is it corrected?

The camera and the slide projector

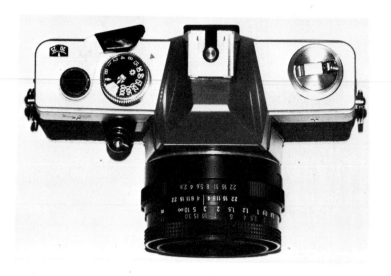

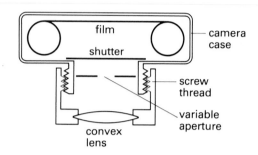

The camera

The camera is an optical instrument that in many ways is similar to the eye. Light enters the camera through the convex lens which focusses the light rays onto the film. The film is similar to the retina in the eye. Instead of changing the shape of the lens, as in the eye, the camera is focussed by altering the distance between the lens and the film. The lens is mounted on a screw thread so that it can be moved in and out.

The amount of light entering the camera is controlled in two ways: by a flap or *shutter* which opens for a certain length of time; and by varying the size of the hole, or *aperture,* through which light enters the camera.

Shutter speed. The time for which the shutter is open is called the shutter speed. On a good camera the shutter speed can be varied between one second and 1/1000 second.

f-stop. Altering the f-stop alters the size of the aperture. The f-stop is made of a series of metal plates which can be moved to increase the aperture size, as shown:

The same amount of light can be let into the camera with a small aperture and long shutter speed, or a large aperture and a short shutter speed. The first combination (small hole, long time) is used for things that are not moving. The small aperture increases the range of distance over which things are in focus. The second combination (large hole, short time) is used to photograph moving objects. Photographs of moving objects often appear blurred because they move during the time that the shutter is open.

The camera in the photograph is a single lens reflex camera, which allows the object to be viewed in exactly the same way as the photograph will come out. The light is reflected by a hinged mirror and a five sided prism into the viewfinder as shown in diagram **1**. The two reflections in the prism are necessary to make the image the right way up. When a photograph is taken, the mirror quickly rises out of the way as shown in diagram **2**:

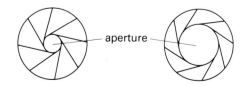

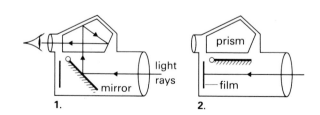

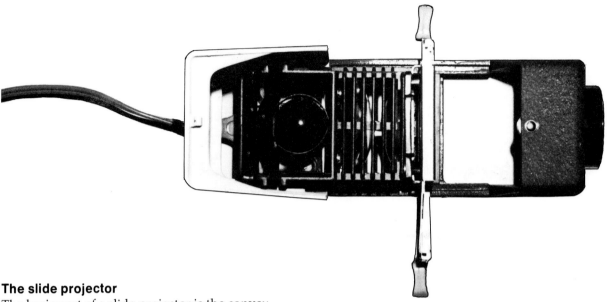

49

The slide projector

The basic part of a slide projector is the convex projector lens which focusses the image of a brightly lit object (the slide) onto a screen. The remaining parts of the projector are to illuminate the slide. All the parts described here can be seen in the diagram, and also in the photograph.

A concave mirror is placed behind the lamp to reflect all the light forward. The condenser lenses are to make the slide evenly illuminated, instead of being brighter in the middle.

The high intensity lamp produces a lot of heat so there is a fan to cool it. The heat filter is a piece of special heat resisting glass placed in front of the slide to stop it getting too hot.

The size of the image on the screen increases as the projector is moved back from it. The image is focussed by altering the distance between the lens and the slide. The projector lens is mounted on a screw thread so that it can be moved in or out to focus the image.

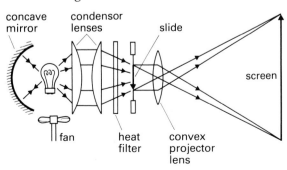

concave mirror / condensor lenses / slide / fan / heat filter / convex projector lens / screen

Exercises

1. Draw a diagram of the camera. Explain how the image is focussed, and how the amount of light reaching the film is controlled.
2. Copy out the following passage filling in the missing words from the lists.

**Comparison of the eye and the camera.
Similarities:**

a. They both have a * lens to * the image onto the * in the case of the eye, and the * in the case of the camera.

b. The amount of light entering is controlled by a variable aperture called the * in the eye and the * in the camera.

(retina, convex, iris, film, stop, focus)

Differences:

a. The eye * by altering the * of the lens. The camera is focussed by altering the * between the * and the film.

b. The eye works * and has no *.

(continuously, shape, shutter, focusses, lens, distance)

3. Draw a labelled diagram of the slide projector. Make a list of the parts and explain what each one does.

Questions on chapter 7

1. The diagram shows two rays of light passing through two lenses and a rectangular glass block.

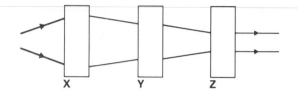

State which part represents the glass block and the type of lens represented by each of the others.

(E.A.E.B.)

2.

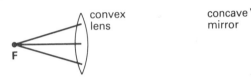

A source of light is placed at **F**, the principal focus of the convex lens, and produces a diverging beam of light incident on the lens.

Copy the diagram and show by drawing further rays:

   i How the beam emerges from the lens,

   ii How the concave mirror affects this emergent beam. (W.J.E.C.)

3.

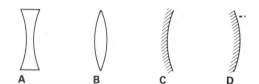

The diagram shows the symbols for four optical items. Using the letters **A**, **B**, **C**, and **D** answer the following questions. (You may have to use more than one letter for each answer.)

  Which (by itself)

   i Can produce images by refraction?

   ii Can produce images by reflection?

   iii Can produce only a virtual image?

   iv Can produce real and virtual images?

   v Can produce magnified virtual images?

   vi Can be used to correct short sight? (S.R.E.B.)

4. a i Why are diverging (convex) mirrors sometimes preferred to plane mirrors for use as car wing mirrors?

   ii Suggest *one* disadvantage of a diverging mirror in this case.

   iii Suggest *one* other place where a diverging (convex) mirror might be used.

  b i Draw a diagram showing how the focal length of a concave mirror may be found using parallel rays of light.

   ii Mark on your diagram the focal length.

   iii How is the focal length of the mirror related to the radius of curvature? (W.M.E.B.)

5. a Images are described as either "real" or "virtual". What do you understand by these terms?

  b You are given three mirrors **E**, **F** and **G**:

   i Name the mirrors.

   ii Which mirror(s) can produce a real image?

   iii Which mirror(s) can produce a virtual image?

  c Copy and complete the following diagrams which show TWO rays of light incident on the mirrors X and Y.

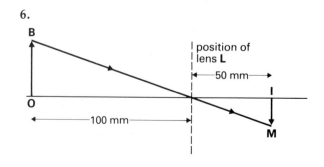

(A.L.S.E.B.)

6.

116

An object **OB**, 30 mm long, is placed in front of a lens **L** such that it gives an image **IM**, 15 mm long, situated 50 mm behind the lens. Copy the diagram to the same scale.

   i Draw the type of lens used.

   ii To locate the image at least two rays of light from the object are required. Complete the diagram by drawing in one more ray from the top of the object **B** to fix the position of its image **M**.

   iii Measure the focal length of the lens.

   iv Describe the image.

   v What would happen to the image if the object was moved further away from the lens?   (E.M.R.E.B.)

7.

1.   2.

The eye in the diagrams above suffers from long sight. Copy them into your book. On diagram **1** draw the rays to show the uncorrected defect. Add a suitable corrective lens to diagram **2** and show the path of the light rays in this case.   (W.M.E.B.)

8. Copy the diagrams into your book.

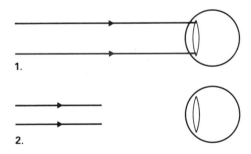

1.

2.

   i Diagram **1** shows two parallel rays of light reaching a short sighted eye. Complete the paths of the two rays on your diagram.

   ii Complete your copy of diagram **2** by showing a lens correcting this defect and then continue the paths of the rays.   (E.A.E.B.)

9. The diagram shows a simple camera:

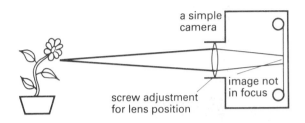

   i Describe two different ways of getting the image into focus.

   ii When the image has been sharply focussed, write down two of its properties.

   iii Explain why some cameras have a variable shutter speed.   (Y.R.E.B.)

10. i Draw a diagram of a simple camera. Show on your diagram the position of the lens, the film, the iris which controls the brightness of the image, and the shutter.

   ii In what two respects is the camera similar to the eye?

   iii In what two respects is the camera different from the eye?

   iv How does the distance between the lens and film change if the focus is altered from a large object distance to a close-up?

   v A special camera, used for photographing postage stamps, has a focal length of 3 cm. It is used to photograph a stamp 3 cm high placed 9 cm from its lens. Redraw and complete the following diagram to show how the image is formed.

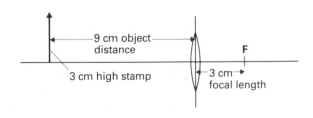

   vi At what distance from the centre of the lens is the image formed?

   vii What is the height of the image?(S.W.E.B.)

117

Waves in water and light

Many different types of energy are transmitted in the form of waves. Light, heat, sound and radio all travel in waves. Water waves release a lot of kinetic energy as they break against a beach or sea wall:

A water wave does transmit energy, but until it "breaks", the water does *not* flow along with the wave. Each part of the water is only moving up and down! This can be shown by placing a float in water waves. The float stays in the same position, bobbing up and down as the *crests* and *troughs* of the wave flow past it:

The movement of the waves is caused by "columns" of water molecules moving smoothly up and down at different times. As some are moving up, others are moving down. The way they move fits into a pattern – the movement of one "column" is always just out of step with the movement of the columns beside it. As a result of this, the water is only moving up and down, but the wave shape is moving forwards. This is shown in the diagram at the top of the next column.

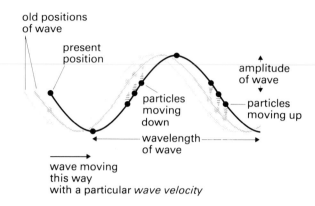

Amplitude, wavelength and wave velocity

Three words are used in describing the movement of waves:

Amplitude. Half the distance that the molecules move up and down is called the *amplitude*. The amplitude of a water wave is shown by half the distance that the float moves up and down.

Wavelength. The distance between the crests of the waves – which is the same as the distance between the troughs of the wave – is called the *wavelength*.

Wave velocity. The speed with which these waves are moving along is called the *wave velocity*.

Transverse waves

In a water wave the water molecules move up and down, but the wave itself moves forwards – the movement of the particles is at right angles to the movement of the wave. This type of wave is called a *transverse wave*:

A transverse wave is one in which the particles vibrate in a direction at right angles to that in which the wave is moving.

Light and water waves are both transverse.

Light waves and water waves

Because waves behave in similar ways to one another, the behaviour of light can be explained by comparison with water waves. Water waves are made in the laboratory in a ripple tank·

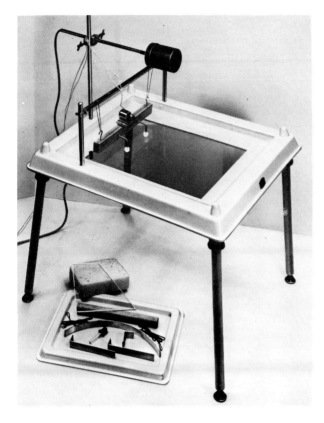

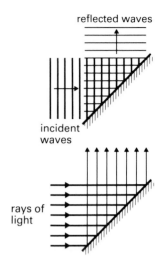

reflected waves

incident waves

rays of light

Refraction at a plane surface. Waves on a ripple tank change direction, or are *refracted,* when the depth of the water changes. A plate of glass is used to make the water shallower. The diagram shows how waves are refracted when they meet the edge of shallower water:

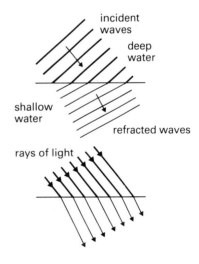

incident waves

deep water

shallow water

refracted waves

rays of light

The wave explanations of some properties of light are suggested below. In each case a photograph of the waves on a ripple tank is on the left of a diagram representing the property. Under this is the normal ray diagram.

Parallel rays of light. These are represented by straight waves travelling in the direction of the rays:

waves moving this way

rays of light

Reflection at a plane mirror. If the waves strike a straight obstacle at any angle, it is found that they are reflected off at the same angle:

Exercises
1. What is a transverse wave?
2. Explain the meaning of the terms: wavelength, amplitude, and wave velocity.
3. Draw diagrams to show how reflection and refraction of light can be explained in terms of the motion of waves.

The electromagnetic spectrum

The electromagnetic spectrum is the name given to a whole series of radiations that include radio-waves, heat radiation, visible light, and X-rays. They are similar in that they all travel at the same velocity, 300 000 km/s. Each type of radiation is different because it has a different wavelength. Radio-waves have a very long wavelength, up to 10 km between crests; and X-rays a very short wavelength, about one millionth of a millimetre between crests.

The complete electromagnetic spectrum is shown in the diagram on the right, together with methods of producing and detecting them.

The visible light spectrum

As can be seen from the diagram, visible light is only a small part of the complete electromagnetic spectrum. The visible light spectrum can be made by passing a beam of white light through a prism:

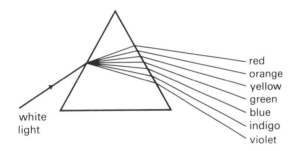

The splitting up of white light into its various colours is called *dispersion*. Dispersion occurs because each colour is refracted by the glass by a different amount. Red is refracted least and violet most.

Each colour of the spectrum has a different wavelength. Violet light has the shortest wavelength – it is closest to the X-ray end. Red light has the longest wavelength, being closest to the radio-wave end.

Red, Orange, Yellow, Green, Blue, Indigo, Violet is the order of spectrum colours – it may be helpful to remember that 'Richard Of York Gained Battles In Vain'.

Recombination of the colours. If a concave mirror is used to bring all the colours of the spectrum to a focus at a single point, an area of white light is formed. White light is made up from a complete mixture of the various colours of the spectrum.

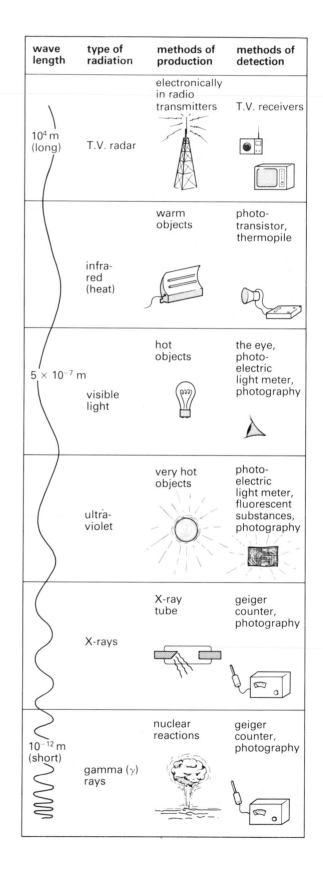

wave length	type of radiation	methods of production	methods of detection
10^4 m (long)	T.V. radar	electronically in radio transmitters	T.V. receivers
	infra-red (heat)	warm objects	photo-transistor, thermopile
5×10^{-7} m	visible light	hot objects	the eye, photo-electric light meter, photography
	ultra-violet	very hot objects	photo-electric light meter, fluorescent substances, photography
	X-rays	X-ray tube	geiger counter, photography
10^{-12} m (short)	gamma (γ) rays	nuclear reactions	geiger counter, photography

51

Ultra violet and infra-red radiation

A spectrum containing these two radiations can be demonstrated in the laboratory using the apparatus shown below. A convex lens is used to focus a strong beam of white light onto a prism:

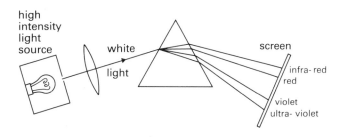

Infra-red radiation. This has a longer wavelength than red light. It is detected by a device called a photo-transistor, which converts the radiation into an electric current:

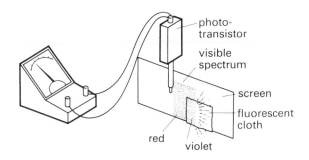

Ultra-violet radiation. This has a shorter wavelength than violet light. A piece of white cloth that has been soaked in a concentrated solution of washing powder and dried, is placed at the violet end of the spectrum. The cloth glows with a white light for a short distance beyond the violet. The ultra-violet light falling on the cloth is being converted into visible light. This effect is called *fluorescence* and is caused by a fluorescent chemical in the washing powder. The fluorescent chemical is added so that white clothes glow in the ultra-violet radiation from the sun making them look "whiter than white".

The rainbow

A person can only see a rainbow when rain is falling in front of him, and the sun is shining from behind him. Every raindrop splits up white light from the sun into the colours of the spectrum, as shown in diagram **1**:

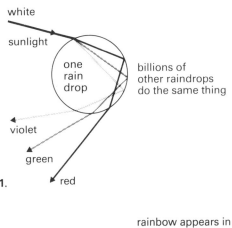

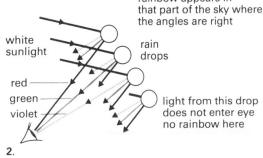

Every rain drop lit by the sun then sends the various colours off in different directions. The eye sees the rainbow by selecting each colour from that set of raindrops in the sky where the angle of entry to the eye is right, as shown in diagram **2**.

Exercises

1. Make a list of the types of radiation in the electromagnetic spectrum in order of increasing wavelength. Label the top of the list "short wavelength", and the bottom of the list "long wavelength".

2. Explain what is meant by *dispersion*.

3. Draw a diagram to show how a prism splits up a beam of white light into the colours of the spectrum. Name the colours on your diagram.

4. Describe how you would detect the presence of infra-red and ultra-violet radiation.

5. Explain how a rainbow is produced. Can you explain why the red is at the top of the rainbow and the violet is at the bottom?

Colour

Each colour of light has its own characteristic wavelength. If light of the yellow wavelength enters the eye, it sees yellow. However, if a mixture of red and green light enters the eye it also sees yellow! It is found that all the colours that the eye sees can be made by mixing three basic colours. These three colours, which are called *primary colours,* are red, blue, and green.

The effect of mixing the primary colours can be demonstrated by using three slide projectors to make circles of red, blue, and green light overlap on a screen as shown:

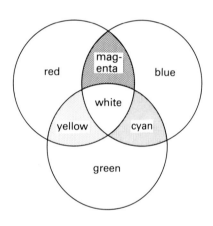

All three colours together make white light. The colours made by mixing any two primary colours are called *secondary colours.* These are magenta (red and blue), cyan (blue and green), and yellow (green and red).

Seeing in colour

Inside the human eye, there are about 126 million light sensitive receptors. They divide into two types: *rods*, which are very sensitive to big variations in light intensity, but only work in black-and-white – and *cones* which work in colour, but only when the light is bright enough. The cones subdivide into three main groups. One set are good at detecting green light, another red light, and the third, blue. It is because of the nature of these cones that the primary colours are green, red and blue. The eye decides which colour it is seeing by comparing the relative amounts of each of these three colours.

Coloured filters

Coloured filters are made out of coloured glass or plastic. A coloured filter transmits (lets through) its own colour, but absorbs any other colour which falls on it. The following examples show how:

White light on a primary colour filter. White light which is made up of red, blue, and green light is shone onto a red filter. The blue and green parts of the white light are absorbed by the filter. Only the red is transmitted making a beam of red light as shown below. Blue and green filters behave similarly in letting only one colour through.

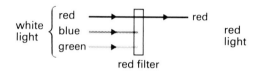

White light on a secondary colour filter. The green part of the white light is absorbed by the magenta (red and blue) filter. The red and blue which make magenta light are transmitted. The secondary colour filter lets two colours through.

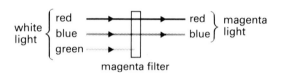

Primary colour filter after a secondary colour filter. The secondary colour filter produces a secondary coloured beam of light. If the primary colour filter is one of the two colours in the beam of light, that colour is transmitted. If the primary colour filter is the colour which is not in the beam of light no light is transmitted. The diagram shows what happens when a red or a green filter is placed after a magenta filter:

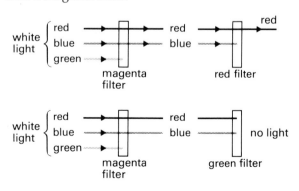

52

Coloured pigments

An object can only be seen when light is reflected from it into the eye. The substance which gives an object its colour is called a pigment. A pigment absorbs all colours except its own, which it reflects.

A black pigment absorbs all colours and reflects none. A white pigment reflects all colours.

The following examples show how coloured pigments work.

White light on a primary colour pigment. The primary colour pigment reflects one colour of light. In the diagram the red pigment is absorbing the blue and green light and reflecting the red:

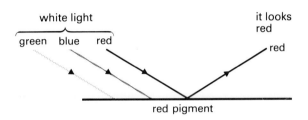

White light on a secondary colour pigment. The secondary colour pigment reflects two colours of light. In the diagram the magenta pigment is reflecting blue and red light but absorbing the green:

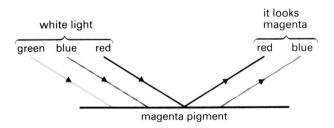

A secondary colour light on a primary colour pigment. If the primary coloured pigment is one of the two colours in the light, that colour is reflected. If the primary colour pigment is the colour which is not in the light, no light is reflected. The diagrams at the top of the next column show what happens when magenta light shines on red and green pigments:

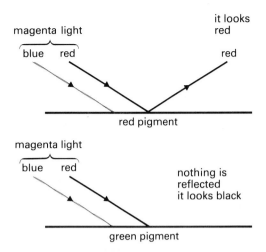

Pigments only appear their true colour when viewed in white sunlight. For example, a white pigment appears yellow when viewed in the light from yellow streetlamps. The light from fluorescent lamps in shops is not exactly the same as sunlight. This is why many women frequently take dress material or clothes to the door of the shop, to check their true colours.

Exercises

1. Explain what is meant by a *primary colour*. Name the three primary colours.

2. Explain what is meant by a secondary colour. Name the three secondary colours.

3. Use diagrams similar to those in the text to show what happens to a beam of white light when it is shone onto:

    a a green filter,

    b a cyan filter,

    c a cyan filter with a blue filter after it,

    d a cyan filter with a red filter after it.

4. Use diagrams similar to those in the text to show what happens when:

    a white light falls on a green pigment,

    b white light falls on a yellow pigment,

    c yellow light falls on a green pigment,

    d yellow light falls on a blue pigment.

Sound waves

The waves that have been considered so far have been transverse waves, in which the direction of vibration is at right angles to the direction of movement of the wave. Sound is a different type of wave called a *longitudinal wave,* in which the direction of vibration is the same as the direction of movement of the wave. Longitudinal waves can be seen in a long coiled spring called a 'slinky'. The spring is stretched and one end is vibrated backwards and forwards. This makes *compressions* (closely spaced coils) and *rarefactions* (widely spaced coils) travel down the spring:

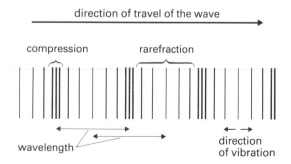

The wavelength of the vibration is the distance between compressions which is the same as the distance between rarefactions. The amplitude depends on the size of the vibration.

A longitudinal wave is one in which the particles are vibrating in the same direction as that in which the wave is travelling.

The properties of the longitudinal wave are summarised in the diagram below:

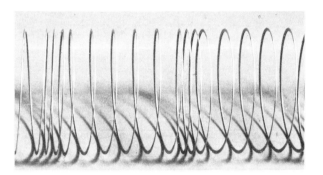

Sound

Sound waves are produced by vibrating objects, and they travel in longitudinal waves. If a rule is held tightly against the edge of a bench, bent, and then released, it vibrates producing a sound. As the rule bends forward it pushes the air molecules together making a compression which moves forward:

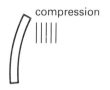

As the rule moves backwards, the air molecules move apart making a rarefaction:

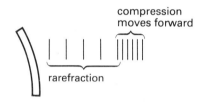

It bends forward again making another compression, which forces the rarefaction and the first compression forwards:

It then bends back again making another rarefaction:

In this way a series of compressions and rarefactions move out from the rule. This is a sound wave.

53

The rarefactions and compressions of the sound wave enter the ear and make a thin, taut piece of skin called the *ear drum* vibrate. This movement of the ear drum is then conveyed to the brain, enabling the sound to be heard.

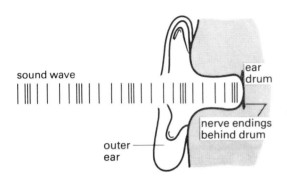

Frequency and pitch

If the free end of the rule is made shorter the rule vibrates faster – the note produced is higher. The rate at which the rule vibrates is called the *frequency*. Frequency is measured in hertz (Hz). 1 Hz is one vibration per second.

How high the note sounds in the ear is called the *pitch*. The pitch of the note is higher when the frequency of the vibration is higher.

The human ear is able to hear frequencies which vary from about 20 Hz (a low pitched hum) to about 20 000 Hz (a very high pitched whistle). Dogs are able to hear frequencies even higher than this.

The loudness of the sound depends on the amount that the source of sound vibrates – the amplitude of the vibration.

The transmission of sound

Sound will travel through any substance that is capable of vibrating:

Solids. The sound of a record player in the next room can be heard through the wall. Sound can travel through solids.

Liquids. The sound made by the propellers of a boat can be heard by a swimmer if he puts his head under the water. Sound can travel through liquids.

Gases. Noise made by aircraft is annoying: sound can travel through gases.

Sound needs a medium. If a vibrating object is not in contact with a solid, a liquid, or a gas, then the sound waves are neither made nor transmitted. This can be demonstrated using the apparatus shown below:

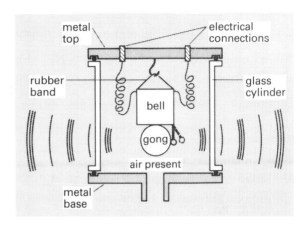

The electric bell is suspended by a rubber band, which does not transmit vibrations; the bell is set ringing. Whilst air is still present the sound of the bell can be clearly heard. When the air is pumped out, the bell can still be seen to be working, but no sound is heard. This shows that sound needs a medium for transmission:

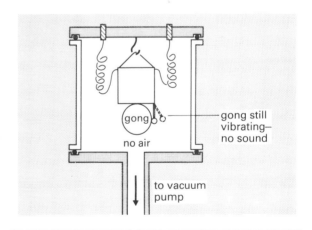

Exercises

1. What is a *longitudinal wave*?

2. Explain the process by which the sound of your voice is transmitted through the air, and is heard by another person.

3. Explain what is meant by the terms "frequency" and "pitch".

4. Describe an experiment which shows that sound needs a medium for transmission.

Velocity of sound

The puff of smoke from a distant starting pistol can be seen before the sound is heard. This can be used to obtain a rough value for the velocity of sound:

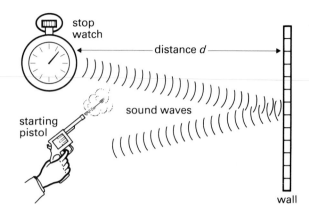

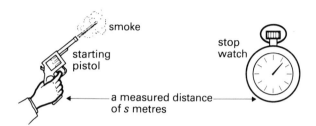

A person with a stop watch measures the length of time between seeing the puff of smoke and hearing the bang of the gun. If the distance from the gun is measured, then the velocity of sound can be found from the formula:

$$\frac{\text{velocity}}{\text{of sound}} = \frac{\text{distance}}{\text{time}} \quad \text{metres/second}$$

$$v = \frac{s}{t} \text{ m/s}$$

Distance of lightning. The same idea can be used to find out how far away lightning is. The time between seeing the lightning and hearing the thunder is measured. Knowing the speed of sound, the distance between you and the lightning can be worked out:

$$\text{distance} = \frac{\text{velocity}}{\text{of sound}} \times \text{time} \quad \text{metres}$$

$$s = v \times t \text{ m}$$

$$\begin{aligned} \text{distance of lightning} &= 333 \times t \quad \text{metres} \\ &= 0.333 \times t \quad \text{kilometres} \\ &= \frac{t}{3} \quad \text{kilometres} \end{aligned}$$

Echoes

An echo is the repetition of a sound, caused by the reflection of sound waves from a hard surface such as a wall or cliff.

The velocity of sound can also be measured from echoes. A person with a starting pistol and a stop watch stands at a distance from a high wall, as shown in the diagram at the top of the next column:

He fires the pistol and measures the time taken for the echo to return. This is the time taken for the sound to travel to the wall and back. The velocity of sound is then worked out:

$$\text{velocity} = \frac{2 \times \text{distance from wall}}{\text{time}}$$

$$v = \frac{2d}{t} \text{ m/s}$$

If the echo of a hand clap is loud enough to be heard, then a more accurate measure can be made. Upon hearing the echo of a first clap, a second clap is immediately given – this is repeated ten times in quick succession. The time for the ten back-forward sound movements is then known.

Echo sounding. Boats frequently carry an instrument called an *echo sounder* which times an echo and calculates the depth of the water, the speed of sound in water being known:

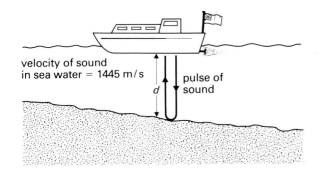

velocity of sound in sea water = 1445 m/s

d

pulse of sound

The frequency of the note depends on the rate at which the compressions (and rarefactions) of the sound wave reach the ear. As the car approaches, it moves closer between making each compression. So the compressions are closer together – the frequency sounds higher.

After the car has gone past, it moves further away between each compression, so there is a greater distance between them. The compressions reach the ear at a slower rate so the frequency sounds lower.

car approaches

compressions closer together

pitch higher

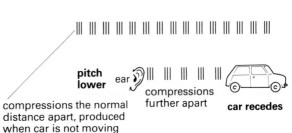

pitch lower

compressions further apart

car recedes

compressions the normal distance apart, produced when car is not moving

The echo sounder measures the time between sending out a pulse and receiving the echo from the sea bed. The depth of the water is worked out from:

$$\text{depth} = \frac{\text{velocity of sound in water} \times \text{time}}{2}$$

$$d = \frac{v \times t}{2} \text{ m}$$

Wave velocity, frequency and wavelength

Sound waves of all frequences travel at the same velocity. If two speakers send out waves at different frequencies, then after one second the *front* wave of each will have travelled the same distance:

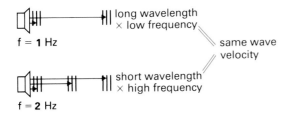

f = **1** Hz

long wavelength × low frequency

same wave velocity

f = **2** Hz

short wavelength × high frequency

This is because the one with a *shorter* wavelength has a higher frequency – the waves are less long, but there are more of them each second. In fact for all waves:

wave velocity = frequency × wavelength

$$v = f \times \lambda$$

The doppler effect

A person standing beside the road whilst a police car drives past sounding its siren will notice that the frequency of the siren appears to change. The frequency of the siren is higher as the car approaches and lower as the car goes away. This effect is called the *Doppler effect*.

Exercises

1. Describe an experiment to measure the velocity of sound in air.

2. A person sees the smoke from a starting pistol 0.9 seconds before he hears the bang. If the pistol is 280 m away, what is the velocity of sound?

3. In a thunderstorm a lightning flash is seen six seconds before the thunder is heard. How far away is the lightning?

4. Explain how an echo sounder works.

5. The echo sounder on a boat sends down a pulse and receives its echo 0.3 seconds later. If the velocity of sound in water is 1445 m/s, how deep is the water?

6. Copy the following table – fill in the missing parts.

wave velocity	frequency	wavelength
322 m/s	Hz	0.5 m
m/s	400 Hz	8.625 m
320 m/s	160 Hz	m

7. What is meant by the term "Doppler effect"?

Vibrating strings and columns of air

All objects, and in particular taut strings such as those on a guitar, will vibrate at one *natural frequency* rather than any other. The frequency of the *fundamental note* produced when a guitar string vibrates at its natural frequency, depends on three things:

Length. The frequency of vibration is higher when the string is shorter.

Thickness. The frequency is higher when the string is thinner.

Tension. The frequency is higher when the string is tighter.

Nodes and antinodes

The vibrating string is held fixed at both ends – at these points the string does not vibrate at all. "No vibration" points are called *nodes*. The string vibrates most at the centre. A "most vibration" point is called an *antinode:*

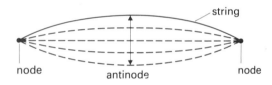

If the string is lightly pressed half way along it, then a node is made there. The string vibrates in two halves, with a higher vibration rate called the *first overtone:*

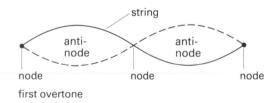

first overtone

A *second overtone* can be made if the string is stopped in two places, one third and two thirds of the way along:

second overtone

Vibrating columns of air

Air columns can also be made to vibrate. Vibrating air is used to make "wind" instruments such as the organ, the flute and the penny whistle. All work on a principle seen the resonance tube:

The resonance tube. This consists of a tube, open at both ends, which stands in a wider container full of water. The length of the air column can be altered by raising or lowering the tube. A U-shaped piece of metal which vibrates at a given frequency, called a *tuning fork,* is set in vibration and placed above the tube as shown in diagram **1**:

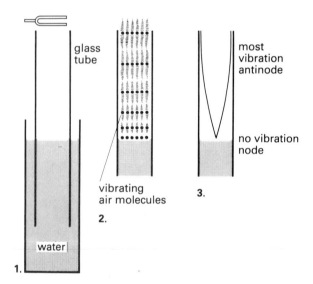

It is found that the sound made by the tuning fork becomes much louder, when the air column is a certain length. The air in the tube now has a natural frequency of vibration which is the same as that of the tuning fork. The vibrations of the tuning fork are setting the air into vibration at its natural frequency. The effect is called *resonance.*

The air is vibrating most at the open end, and is not moving at all in the closed end, as shown in diagram **2**. There is a *stationary wave* a quarter of a wavelength long in the tube. As the longitudinal waves are difficult to draw, all the stationary waves in pipes are drawn as transverse waves as in diagram **3**, where the *width* of the wave suggests the amount of up-and-down movement.

55

Closed organ pipes. When an organ pipe is blown gently the passage of air through the slot sets the air into vibration at its natural frequency. The end where the slot is forms an open end – the air molecules vibrate most there:

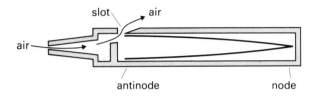

If the organ pipe is blown harder, higher notes are obtained. These are the overtones. The air is always set into vibration in such a way that there is an antinode (most movement) at the open end and a node (no movement) at the closed end:

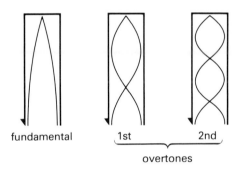

fundamental   1st   2nd

overtones

Open organ pipes. The stationary waves set up in organ pipes that are open at both ends have antinodes at both ends:

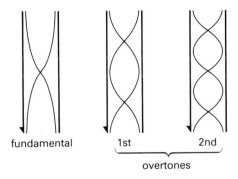

fundamental   1st   2nd

overtones

Frequency of notes from organ pipes. The frequency is higher when the pipe is shorter. An open pipe will produce notes at half the frequency of a closed pipe of the same length.

Waveforms

Sound can be investigated with a *cathode ray oscilloscope,* which shows the compressions and rarefactions of sound, as a picture or *waveform.*

Diagram **1** shows the waveforms of sounds of the same frequency but different loudness. Diagram **2** shows waveforms of two sounds of the same loudness but different frequency:

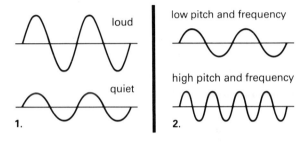

The sounds shown by the waveforms above are those of pure notes. A pure note sounds rather flat and dull. The quality of a note comes from faint overtones being sounded along with the fundamental note. The waveforms below show how the fundamental and an overtone will combine to form a "richer" note:

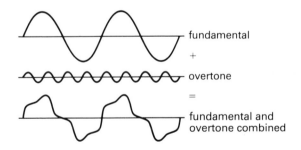

Exercises

1. Draw diagrams to show how a stretched string can vibrate at its fundamental frequency and also at its first two overtones.
2. Explain three ways in which the note from a stretched string can be made higher.
3. Draw diagrams to show how the air in organ pipes (a) closed at one end, and (b) open at both ends, vibrates with the fundamental frequency and the first three overtones.
4. Draw the wave forms which are produced by two notes of (a) the same pitch but different loudness, (b) the same loudness but different pitch, and (c) different quality.

Questions on chapter 8

1.

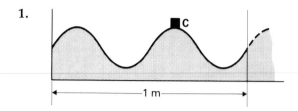

The diagram shows a long tank of water in which waves are produced. **C** is a small cork.

    i 4 complete waves are produced every second. What is the frequency of the waves?
    ii What is the wavelength?
    iii How does the cork **C** move as the wave progresses?   (S.R.E.B.)

2. a Radio 4 broadcasts on 1 500 m.
    i To what does 1 500 m refer?
    ii If radio waves travel at 300 million metres/second, what is the frequency of a wave of wavelength 1 500 metres?
    b Give the correct order in **increasing** wavelength, for the following waves: infra-red, visible, gamma, X-ray, ultra-violet, radio.
    c Describe how the pitch of a horn changes when a car sounding its horn passes you on the road. Give the name of this effect.
    d A man is standing 85 metres from a cliff face in front and 170 metres from a cliff face at his back. He blows a whistle loudly and hears two echos, one after the other.
Explain how this can happen and calculate how long the first echo took to reach the man after he blew the whistle.
(The velocity of sound in air is 340 metres/second).

                              (E.A.E.B.)

3. A beam of white light passes through a triangular glass prism onto a white screen as shown in the diagram:

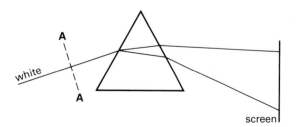

i Copy the diagram and write on it, in the correct position, the names of any colours which would appear on the screen.
ii What would be the effect of placing a magenta filter at **A-A**?   (E.A.E.B.)

4.

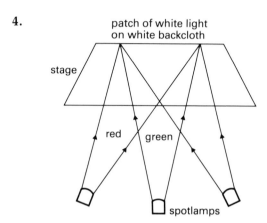

A stage is illuminated by three spotlight beams. One of them is red and one is green.
    i If the three beams together produce a patch of white light on the back cloth, what is the colour of the third beam?
    ii What general name is given to this group of colours?
    iii A dancer in a yellow dress comes onto the stage. Which spotlamp beam or beams should be directed on her to make the robe appear red?
    iv How can her robe be made to appear black without making the rest of her invisible?
    v Which beam or beams should be used to make the backcloth appear yellow?   (Y.R.E.B.)

5. A stage electrician has available red, blue and green lights.
    i What colours must he shine onto a white wall to obtain yellow?
    ii If he shines all three colours onto the same part of the wall, what colour will he obtain?
    iii If he shines a red light onto a blue painted wall, what will be the appearance of the wall?   (W.M.E.B.)

6. i What conditions are necessary for an observer to be able to see a rainbow?
ii Explain why a rainbow has many colours in it.
iii What would be seen if a rainbow were looked at through a red filter? Give a reason for your answer. (M.R.E.B.)

7.

A   **B**

a The diagrams above represent two different types of wave motions carrying energy in the directions shown. Name the wave motion represented by
   i diagram **A** and
   ii diagram **B**.
Explain the main difference between these two types of wave motion.
b State which of the above waves carries energy through
   i air,
   ii A stretched string when plucked,
   iii A metal bar when tapped. (W.J.E.C.)

8. i What is an echo?
ii What two factors affect the length of time taken for an echo to be heard?
iii A ship, depth sounding, receives a weak echo from a shoal of fish three seconds before a stronger echo from the sea bed. If the velocity of sound in sea water is 1 500 ms⁻¹ how far is the shoal of fish from the sea bed? (N.W.R.E.B.)

9. The diagram shows a stretched string vibrating at its fundamental frequency.

   i Copy this diagram and draw two others to show the wire vibrating at its next two higher possible frequencies.

ii State two factors which affect the frequency of the fundamental note produced by the string.
iii Why does the quality (or timbre) of the sound produced by a violin differ from that produced by a piano? (W.M.E.B.)

10. One end of a string is moved to and fro so that a transverse travelling wave is produced.

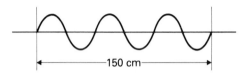

   i What is the wavelength?
   ii If it moved to and fro 2 times a second, what is the frequency of the wave?
   iii What is the velocity of the wave?
   (A.L.S.E.B.)

11. What effect on the frequency of the note emitted by an organ pipe closed by a stop at one end, is achieved by
   i Narrowing the pipe?
   ii Lengthening the pipe?
   iii Removing the stop? (N.W.R.E.B.)

12.

  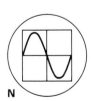

L   **M**   **N**

The diagram shows the trace on a cathode ray oscilloscope screen obtained when three notes are sounded in succession in front of a microphone connected to the oscilloscope. (the controls of the oscilloscope are not altered during the experiment.)
   i In what way (or ways) does note **M** differ from note **L**?
   ii In what way (or ways) does note **N** differ from note **L**? (W.M.E.B.)

Static electricity

If a plastic pen is rubbed with a cloth it will attract small pieces of paper. As a nylon pullover is removed crackling can be heard – if the room is dark, sparks can be seen, too. These effects are due to *static electricity*.

Theory of static electricity

To understand static electricity effects it is necessary to know something about the structure of the atom. It is made up from *protons, neutrons,* and *electrons.* The mass and positive charge of the atom is concentrated in a small volume, called the *nucleus,* at the centre. The nucleus contains protons and neutrons. Around the nucleus revolve the very small negatively charged electrons. The electrons have a negative charge, and the protons a positive charge – the total negative charge on the electrons is equal to the total positive charge on the nucleus:

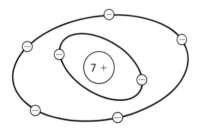

The nuclei of the atoms in a solid are fixed in a regular pattern, so the positively charged protons cannot move – only the electrons are able to move. Substances through which electrons can move easily are called *conductors.* Substances through which electrons find if difficult to move are called *insulators.* In general, all metals are conductors, and non-metals are insulators.

All objects normally contain electrons and protons in equal numbers – they are *uncharged.* An object which is electron rich is *negatively charged* and an object which is electron short is *positively charged.*

The cover photograph shows a machine designed to build up very large static electricity charges. The rear cover shows old-fashioned equipment, used in early experiments on static electricity.

Laws of electrostatics

These laws describe the behaviour of charged objects when they are brought near to one another. If a piece of polythene is rubbed against a clear plastic rule made of acetate, the polythene pulls electrons off the acetate:

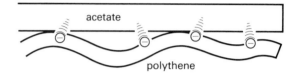

This method of charging is called *charging by friction.* The polythene becomes electron rich (negatively charged) and the acetate becomes electron short (positively charged):

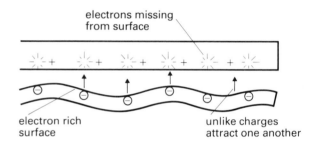

This makes a force of attraction between the acetate and the polythene. If they are separated they pull together. This is the first law of electrostatics:

unlike charges attract one another.

It is sometimes convenient to charge by friction by rubbing rods of the substance concerned with a cloth. A polythene rod gains electrons from the cloth, acetate loses electrons to it. If a charged acetate rod is suspended by a nylon thread, it is repelled by another charged acetate rod. Charged polythene rods give the same result – this is the second law of electrostatics:

like charges repel one another.

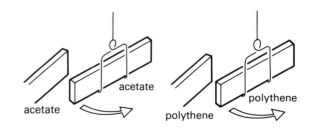

56

Charging by induction

This method of charging conductors uses a charged rod of an insulator. Two metal spheres on insulating stands are placed with the spheres touching. A negatively charged rod is brought close to one of the spheres. This pushes electrons from the near sphere onto the further one, leaving the near one positively charged:

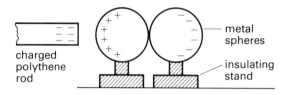

The two spheres are then moved apart:

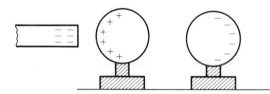

The charged rod is removed and the two spheres are left charged. One has a positive charge and the other has a negative charge:

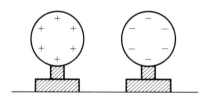

The spheres could also have been charged by induction using a positive rod – only the sphere on the left would have a negative charge, and on the right, positive.

The gold leaf electroscope

This instrument is used to detect charge, and to measure quantity of charge. The main part of it is the metal cap and stem. Mounted against the bottom of the stem is a piece of very fine, flexible gold leaf:

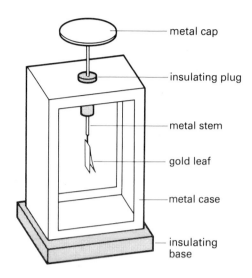

When a charged rod *comes near* the cap, the leaf rises. In a similar way to the spheres, the same charge is induced at the furthest away point from the rod – this is the stem and the gold leaf. Both of them have the same charge – there is repulsion between them, and this causes the leaf to rise. The leaf will fall when the rod is removed.

When a charged rod *touches* the cap, charge is transferred to the electroscope, and the leaf will stay up after the rod is removed.

Exercises

1. Copy out the following passage using words from the list.

All materials contain positive and negative charge in * amounts. The * charge is on the nuclei of the atoms, which cannot move. The * charge is on the electrons. The electrons can pass freely through a * but have great difficulty in moving through an *.

(insulator, positive, conductor, negative, equal).

2. State the laws of electrostatics.

3. Explain with the aid of diagrams how to charge two metal spheres by induction using a *positively* charged rod.

4. Draw a labelled diagram of a gold leaf electroscope with a positive rod near it.

Electric cells

Electric cells produce electricity from a chemical reaction. All cells consist of two plates of different substances placed in a liquid. The plates which are called *electrodes* are made from a metal or carbon. The liquid, which is called an *electrolyte* is a solution of an acid, a salt, or an alkali in water.

The chemical reaction between the electrodes and the electrolyte makes one electrode electron rich (negative) and the other electron short (positive). If the two electrodes are connected by a piece of wire electrons flow from the negative electrode to the positive electrode:

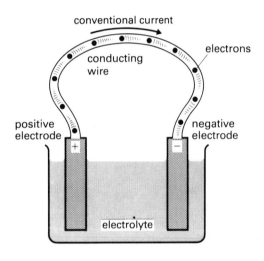

Electric cells are divided into two types: primary cells and secondary cells.

Primary cells
In these cells the chemical reaction which produces the electric current is irreversible. These cells cannot be recharged. The simple cell, and the dry cell are both primary cells.

Simple cell. This is the basic cell from which all others have been developed. It consists of a zinc plate and a copper plate in dilute sulphuric acid, as shown in the diagram at the top of the next column:

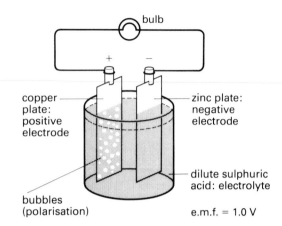

When a bulb is connected between the two plates a current flows. The zinc dissolves making the zinc plate electron rich, and hydrogen is given off at the copper plate making it electron short. Electrons therefore flow through the bulb, causing it to light. The formation of hydrogen bubbles on the copper plate stops the current flowing and is called *polarization*. Polarization makes the brightness of bulb fade after about a minute.

Dry cell. This cell uses carbon as the positive electrode, zinc as the negative electrode and a paste of ammonium chloride as the electrolyte:

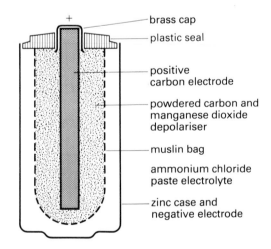

Polarization is prevented by surrounding the carbon rod with a muslin bag containing powdered carbon and manganese dioxide. This mixture acts as a depolorizer by absorbing the bubbles of hydrogen.

57

Secondary cells

This type of cell can be recharged. A chemical reaction takes place, producing electricity in the same way as in the primary cell. The chemical reaction can be reversed, by passing a current through the cell in the opposite direction to the current obtained when the cell was being used. Two common types of secondary cells are the lead-acid accumulator, and the nickel-cadmium cell.

Lead-acid accumulator. The diagram shows a simple lead-acid accumulator used for laboratory demonstration:

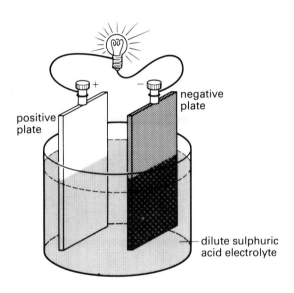

When the cell is charged the positive electrode is coated with lead oxide and the negative electrode is pure lead. As the cell is used both plates react with the dilute sulphuric acid electrolyte producing water and coating the plates with lead sulphate.

The water which is formed makes the sulphuric acid more dilute and less dense. Therefore measuring the density of the electrolyte is a way of measuring the amount of charge in the cell.

In practice the lead acid cell does not consist of just two plates, but two sets of plates connected together as shown in the diagram at the top of the next column. Batteries of this type are used in cars and large motor cycles:

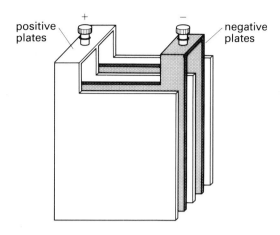

Lead-acid accumulators need regular attention. They should be charged once a month whether they have been discharged or not. Water from the electrolyte slowly evaporates and this should be replaced with distilled water.

Nickel-cadmium cell. These cells are made to the same shape and size as dry cells. They are used for powering electric calculators, razors and torches:

Exercises
1. Explain the difference between primary and secondary cells.
2. Explain what is meant by *polarization* and how it is overcome in the dry cell.
3. How is it possible to measure the amount of charge in a lead acid-cell by measuring the density of the electrolyte?
4. Copy out the following sentence filling in the missing words from the list

   The lead-acid * should be * at least once a month, and any water that has * from the * should be replaced with * water.

(charged, accumulator, electrolyte, distilled, evaporated)

Electrolysis

The electrical conductivity of a liquid can be investigated by dipping two carbon electrodes into it, and seeing if it will pass a current.

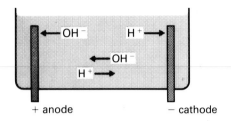

+ anode   − cathode

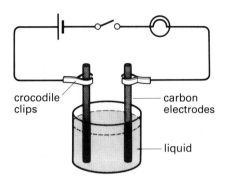

Some liquids such as paraffin, alcohol, and oil do not conduct electricity so the bulb does not light up. Solutions of acids, alkalis, and salts in water do conduct electricity. In some cases gas bubbles are formed at the electrodes or new substances are deposited on them. The solution is being broken down or decomposed by the current.

Any liquid which conducts an electric current and is decomposed by it is called an *electrolyte*. The process is called *electrolysis*.

The electrolysis of water
The molecules of electrolytes conduct electricity because they can split up into charged particles called *ions*. In the case of water some of the molecules are split up into ions as shown by the equation:

$$H_2O \rightarrow H^+ + OH^-$$

water molecule   positive hydrogen ion   negative hydroxyl ion

One hydrogen atom from the molecule has lost an electron to become a positive ion. The remaining hydrogen atom and the oxygen atom collect the electron to become a negatively charged *hydroxyl* ion.

In the electrolysis cell the positive ions are attracted to the negative electrode (the cathode) and the negative ions are attracted to the positive electrode (the anode):

The hydrogen ions collect electrons from the cathode to replace the ones that they have lost, and bubbles of hydrogen gas are released. At the anode the hydroxyl ions give up their extra electron. Two of them then react to form water and oxygen-bubbles of oxygen are released.

The electrolysis of water is carried out in the laboratory using the apparatus shown. It is found that the experiment works better if a small quantity of sulphuric acid is added to the water.

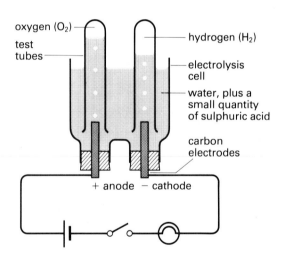

Electrolysis of copper sulphate using copper electrodes
The copper sulphate splits up into ions as shown by the equation:

$$CuSO_4 \rightarrow Cu^{2+} + SO_4^{2-}$$

                  copper ion   sulphate ion

The water is also split up into ions.

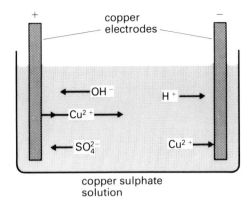

copper electrodes

+ −

OH⁻

H^+

Cu^{2+}

SO_4^{2-}

Cu^{2+}

copper sulphate
solution

In the electrolysis cell both the copper and hydrogen ions are attracted to the cathode. Copper is deposited in preference to hydrogen being released. Both sulphate and hydroxyl ions are attracted to the anode, but copper goes into solution in preference to either being released. The net result is that copper is transferred from the anode to the cathode! This process is called *electroplating*.

Electroplating is used to purify copper. The anode is made of impure copper and the cathode is made of a thin sheet of pure copper. Copper dissolves off the anode and is plated onto the cathode. The impurities in the anode sink to the bottom of the tank. The photograph shows impure sheets of copper being loaded into an electrolysis cell in a copper refining plant in Zambia, Southern Africa.

Electroplating is frequently used to deposit a thin layer of metal onto metal objects to either make them look attractive – silver plating, or to protect them from rust – chromium plating. In the electrolysis cell the anode is made of the metal to be transferred. The electrolyte is a solution of a salt of that metal in water. The objects which are to be coated with the metal form the cathode.

Extraction of metals by electrolysis

As electrolysis decomposes substances, it is used to extract metals. Aluminium is produced in this way. An electric current is passed through a molten mixture of aluminium ores and molten aluminium is released at the cathode. The process uses a huge amount of electrical energy.

Exercises
1. Draw a diagram of the apparatus that you would use to investigate the electrical conductivity of a liquid. State what kind of solutions conduct electricity.
2. What is an electrolyte?
3. Draw a diagram of the apparatus used for the electrolysis of water. Label the anode, the cathode and the places where the oxygen and hydrogen are released.
4. You are given a piece of silver, a beaker containing silver nitrate solution, a battery, and connecting wires. Draw a diagram to show how you would use this apparatus to silver plate a metal spoon. Mark on your diagram the anode and cathode.

Electrical circuits

An electric cell has two terminals. The chemical reaction in the cell makes the negative terminal electron rich and the positive terminal electron short. If the terminals are joined by a piece of conducting wire to make a *circuit,* then electrons flow between them. The electrons flow from the negative terminal to the positive terminal. This is called the *electron current:*

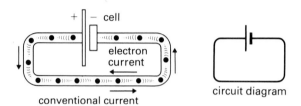

conventional current

circuit diagram

A *conventional current* which is thought of as flowing from the positive to the negative has been used for many years. Since it flows from the positive to the negative:

The conventional current flows in the opposite direction to the electron current.

Ohms and amps

The electrons flowing in the above circuit are doing no useful job. The only effect is to make the cell get hot, and go flat.

A light bulb glows if it is placed in a circuit. The part that glows is called the *filament.* The filament is made of a piece of *resistance wire,* which does two things. First, it reduces the rate of flow of electrons. Second, electrons lose some energy as they pass through the resistance wire. This energy is converted into heat making the filament glow white hot. The resistance of a wire or bulb is measured in *ohms* (symbol Ω)

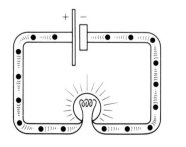

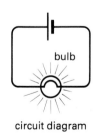

bulb

circuit diagram

A flow of electrons round the circuit is called a *current.* The size of the current depends both on the "push" of the battery, and the resistance of the circuit. Current is measured in *amperes,* or *amps* for short (symbol A). The size of the current is measured with an ammeter.

Using the ammeter

The ammeter must be placed in the circuit to measure the rate of flow of electrons:

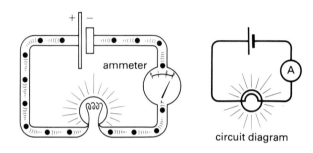

ammeter

circuit diagram

Since the electrons flow all the way round the circuit, the current is the same all the way round it. The ammeter can therefore be placed at any position in the circuit and the reading will be the same:

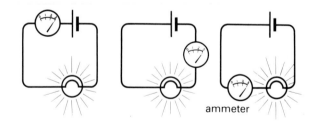

ammeter

Electromotive force

The "push" of the battery that drives the electric current round the circuit is called its *electromotive force* (E.M.F.). The electromotive force of one cell is usually big enough to make a current sufficiently large to light one bulb brightly. It will not have enough force to drive the same current through the higher resistance of two bulbs one after the other – in *series.* Two cells in series would be needed to provide twice the electromotive force, as shown in the next diagram:

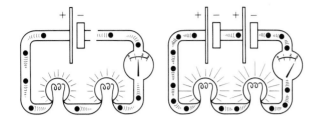

59

Any number of cells can be connected in series and the total electromotive force is equal to the sum of the individual electromotive forces.

Another way of connecting cells is to put them alongside one another, or in *parallel*. In this case the total electromotive force is the same as that of one cell. The advantage of this arrangement is that it will give a larger current.

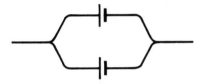

Volts
The part of the electromotive force that drives the current across one particular bulb or resistance in the circuit is called the *potential difference* across that bulb. Potential difference is measured in *volts* (symbol V). A voltmeter measures potential difference.

Using the voltmeter
The voltmeter is *not* part of the main circuit. To measure the potential difference between two points, the voltmeter is fitted *alongside* the main circuit, and connected to the two points, as in diagram **1** below:

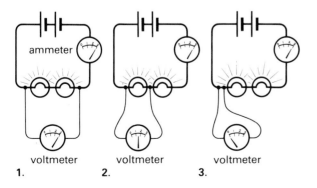

The resistance of the two bulbs splits up the electromotive force of the cells between them. When the voltmeter is connected across both bulbs, as in diagram **1**, it reads a potential difference of 3 V. When the voltmeter is connected across one bulb it reads 1.5 V, as in diagram **2**. As the connecting wires have almost no resistance almost no potential difference is needed to drive the current through them. The potential difference across any two positions on the same piece of connecting wire is zero, as in diagram **3**.

Bulbs in parallel
One cell provides enough electromotive force to light one bulb brightly. Two bulbs in series split the electromotive force of the cell between them. The potential difference across each of the bulbs is then too small to drive enough current through either bulb to make it glow brightly. If the two bulbs are connected in parallel with the cell, each bulb has the full potential difference across it. As long as the battery is in good condition, enough current will flow through each bulb to make it glow brightly:

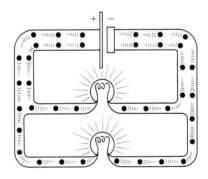

The life of the battery using it this way, will be halved.

Exercises
1. Explain what is meant by *electric current* and *potential difference*.
2. In what units are current, potential difference, and resistance measured?
3. Draw a circuit diagram to show how a bulb can be lit from an electric cell. Include in your circuit an ammeter to measure the current, and a voltmeter to measure the potential difference across the bulb.

Electrical resistance

Some things have a high resistance to the passage of electricity through them, and others have a low resistance. The resistance of any object is fixed but it is difficult to measure directly. The resistance of an object is found by measuring the current flowing through it when a known potential difference is used.

An object has a resistance of 1Ω if a potential difference of 1 V causes a current of 1 A to flow.

Ohm's law
This law concerns what happens when the potential difference across a resistor increases, as in the following diagrams:

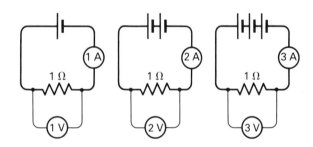

It is found that a potential difference of:
1 V drives 1 A through the 1 Ω resistor;
2 V drives 2 A through the 1 Ω resistor;
3 V drives 3 A through the 1 Ω resistor.

From this it can be seen that:

$$\frac{\text{potential difference}}{\text{current}} = \text{resistance}$$

$$\frac{V}{I} = R$$

The results show that the current increases in step with the potential difference. In other words, the current is *proportional* to the potential difference. This statement is *Ohm's law*:

Provided the temperature remains constant, the current flowing through a resistor is proportional to the potential difference across its ends.

The condition "provided the temperature remains constant" must be included because the resistance of metals increases as they get hotter.

The effect of shape on resistance
The resistance of an object depends on its length and thickness:

Resistance is higher for a longer wire. When the wire is longer there are more obstacles to the flow of the electric current, so the resistance is higher.

Resistance is higher for a thinner wire. When the wire is thinner there are fewer paths for the current to flow through so the resistance is higher.

The variable resistor. This consists of a coil of resistance wire with a sliding contact that can be moved along the wire:

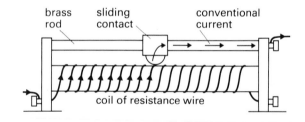

As the sliding contact is moved to the right, the current has to flow through a greater length of resistance wire. This increases the resistance and reduces the current.

60

Resistors in series. When the resistors are connected in series as shown, the length of resistance wire that the current passes through is increased:

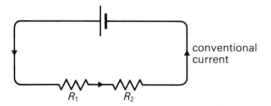

The resistance to the flow of current is increased. The effective resistance is found by adding the two resistances together:

$$R_{\text{total}} = R_1 + R_2$$

Resistors in parallel. When two resistors are connected in parallel, as shown below, the current has two paths to flow through instead of one. So the resistance to the flow of current is *reduced:*

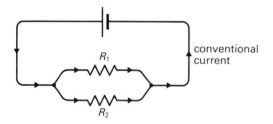

The resistance of two resistors in parallel is lower than either of them by itself, and can be worked out from the formula:

$$\frac{1}{R_{\text{total}}} = \frac{1}{R_1} + \frac{1}{R_2}$$

Current flowing through resistors in parallel. When *equal* resistors are connected in parallel the current divides exactly in two, so that half flows through each resistor:

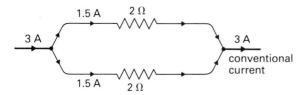

When one resistor is larger than the other, more current takes the easier passage – through the lower resistor. In the example below 2 A flows through the 1 Ω resistor and 1 A throught the 2 Ω resistor:

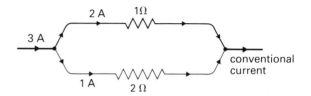

Example. Two resistors of 2 Ω and 4 Ω are connected in parallel. What is the effective resistance?

In parallel, the total resistance is given by:

$$\frac{1}{R_{\text{total}}} = \frac{1}{R_1} + \frac{1}{R_2}$$

$$= \frac{1}{2} + \frac{1}{4}$$

$$\frac{1}{R_{\text{total}}} = \frac{2+1}{4} = \frac{3}{4}$$

$$\therefore R_{\text{total}} = \frac{4}{3} = 1.33 \ \Omega$$

$R_{\text{total}} = 1.33 \ \Omega$, i.e. *less* than either of the two resistors by themselves.

Exercises
1. State Ohm's law.
2. What effect does the length, diameter, and temperature of a metal wire have on its resistance?
3. Draw a circuit diagram to show how you would measure the resistance of a light bulb using a dry cell and an ammeter and voltmeter. If the reading of the voltmeter was 1.5 V and the reading of the ammeter was 0.3 A, what was the resistance of the bulb?
4. A resistor of 4 Ω has a potential difference of 2 V across it; what current flows?
5. What is the effective resistance of a 4 Ω resistor and a 2 Ω resistor connected in series? If a current of 2 A flows through them, what is the potential difference across each resistor?
6. What is the effective resistance of two 2 Ω resistors placed in parallel?

Electrical heating

Heat energy is produced whenever an electric current flows through a wire. The heating depends on the current flowing and the resistance of the wire. In an electric fire, the connecting cable has a very low resistance but the wire of the heating element has a high resistance so virtually all the heat is produced at the element.

Theory of electrical heating.
As the electrons of the electric current flow down the metal wire they knock into the nuclei of the metal atoms:

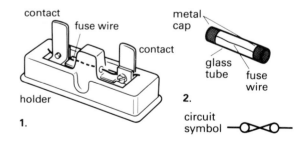

In each collision the electrons lose some of their kinetic energy – they give it to the nuclei of the atoms making them vibrate faster. This vibration of the atoms shows itself as heat – the wire gets hotter. The total amount of heat produced depends on:

1. Resistance. If the resistance is higher there are more obstacles for the electrons to collide with.

2. Current. If the current is higher there are more electrons to collide with the obstacles. Experiment shows that the amount of heat produced depends on the square of the current (I^2).

3. Time. If the current flows for a longer time more collisions occur.

These three effects are summarized in the equation:

heat produced = resistance × current² × time
(joules)   (ohms)   (amps)   (seconds)

$$H = R I^2 t$$

Electrical power
Because the resistance can be worked out from: $R = V/I$, the heat produced equation can be re-written in another form:

$$\text{heat produced} = \frac{\text{potential}}{\text{difference}} \times \text{current} \times \text{time}$$

$$H = VIt$$

This form of the equation is more convenient because the potential difference can be measured with a voltmeter and the current with an ammeter.

Dividing both sides of the equation by t gives:

$$\frac{H}{t} = VI$$

H/t indicates the rate at which heat energy is produced. This is the *power* being used:

power   = potential difference × current
(watts)   (volts)   (amps)

$$P = VI$$

Fuses
A fuse is a short piece of thin copper wire which gets hot as the current flows through it. Too large a current going through it makes it melt. This breaks the circuit and switches off the current.

Fuses are used to protect all electrical appliances in the home. Diagram **1** below shows the type that is used in the main fuse box of the house. A melted piece of fuse wire can be replaced by fastening a new piece between the contracts. Diagram **2** shows a cartridge-type fuse. When the wire has melted in this fuse, the whole cartridge has to be replaced:

The power sockets in most houses are now of the "square pin" type – each plug is fitted with its own fuse, as shown in the next diagram.

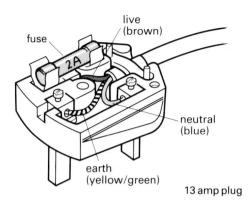

fuse

live (brown)

2A

neutral (blue)

earth (yellow/green)

13 amp plug

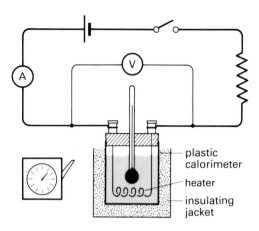

plastic calorimeter

heater

insulating jacket

The fuse is connected between the live pin of the plug and the live wire of the appliance so that if it blows, the live wire is disconnected. The three wires of the flexible cable to the plug are colour coded so that the correct connections can be made. The colour code is:

Live wire (marked **L** on plug) brown
Neutral wire (marked **N**) blue
Earth wire (marked **E**) yellow and green

In old wiring, the live is red, the neutral is black, and the earth is green.

Choice of fuse. An appliance is designed to take a certain current. Fuses are made in certain standard values. The fuse fitted in the plug should be the one whose current rating is just above the current taken by the appliance.

The problem is that the appliance usually has the power marked on it, and not the current. As the potential difference of the mains is 240 V, the current may be worked out from the form of the power equation:

$$\text{current} = \frac{\text{power}}{\text{potential difference}}$$

$$\text{current} = \frac{P \text{ watts}}{240 \text{ V}}$$

Specific heat capacity of a liquid by an electrical method.

Electrical heating is a convenient method of measuring specific heat capacity (see page 60.) An electric current is passed through a heater immersed in a mass of liquid m at a temperature of θ_1. After a time t the new temperature θ_2 is measured. The potential difference across the heater and the current flowing are measured with the voltmeter and ammeter.

$$\text{heat gained by liquid} = \text{electrical energy supplied}$$

But:

$$\text{heat gained by liquid} = \text{mass of liquid} \times \text{specific heat} \times \text{temp rise}$$

$$= m c (\theta_2 - \theta_1)$$

$$\text{electrical energy supplied} = \text{potential difference} \times \text{current} \times \text{time}$$

$$= V I t$$

Putting these together;

$$m c (\theta_2 - \theta_1) = V I t$$

$$c = \frac{V I t}{m (\theta_2 - \theta_1)}$$

Exercises

1. Copy out this passage on the theory of electrical heating using the words below it.

Theory of electrical heating.
As the * flow down the wire past the * of the atoms, they knock into them. In each * the electron loses some of its * to the nucleus, which * faster, making the wire hotter.
(energy, nuclei, collision, electron, vibrates)

2. Mains fuses are made in values of 2, 3, 5, 7, 10, and 13 amps. What fuse would you choose for each of: a 750 W electric fire, a 2000 W electric radiator, and a table lamp with a 60 W bulb.

3. When a potential difference of 5 V is applied across a heater in a calorimeter a current of 2 A flows. If this raises the temperature of 0.1 kg of methylated spirit from 17 °C to 27 °C in 230 s, what is the specific heat capacity of the methylated spirit?

Electricity in the house

The electrical supply in the United Kingdom has a potential difference of 240 V. It enters the home through an installation similar to the one shown.

The power is connected first to the electricity board's fuse box. This is sealed and must not be opened by anybody other than an electricity board official. The power is supplied by means of two cables. One cable, coloured brown, is called the *live* and is at 240 V. The other cable, coloured blue, is called the *neutral*. This is connected to the Earth at the power station and like the Earth, is at 0 V. (In old wiring, the live is red and the neutral is black.)

From the electricity board's fuse box the current passes through the meter to the domestic fuse and distribution box. Here the electricity is directed to electrical equipment such as the cooker, power sockets, water heater, and lights. Each circuit has its own fuse which is connected in the live wire.

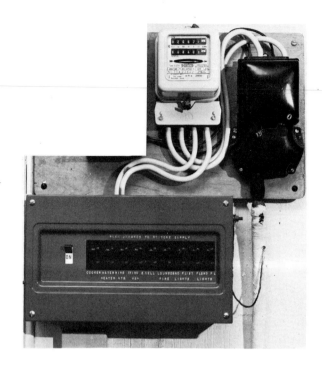

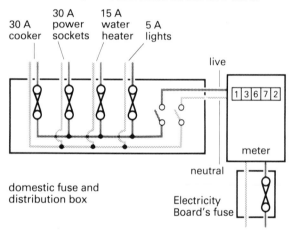

domestic fuse and distribution box

Electricity Board's fuse

The lighting circuit
The diagram in the next column shows how the lights are connected. Each bulb is placed in parallel with the main supply so that there is a potential difference of 240 V across each bulb. The switch is always placed in the live wire so that, when the light is switched off, the bulb and the wires past the switch are at 0 V.

Switch **1** controls only one bulb. Switch **2** controls two bulbs – both go on or off at the same time. The bottom two bulbs are connected up to "two-way" switches. A two-way switch transfers the current from one route to the other. Switching either one will cause the bulb to go either on or off. Two-way circuits are often used at the top and bottom of stairs, to control the bulb that lights them.

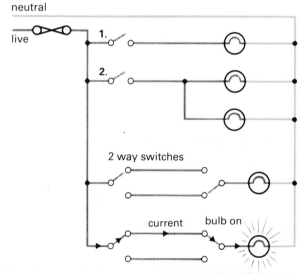

Power sockets and the earth connection
Each power socket in the house has three connections. The third connection is called the *earth*. The earth socket is usually connected to a metal water pipe which makes good contact with the ground of the Earth. The neutral wire is connected to the Earth at the power station. The earth socket provides another path for the electric current to return to the dynamo.

62

The earth wire. This is a safety precaution which is connected to any metal parts of the case of an appliance. If a bare live wire touches a metal case which is *not* earthed, the case becomes live and at 240 V. If a person touched this case the electric current would flow to Earth *through him* – he would receive a severe electric shock. If the live wire touches a case which is earthed, current flows through *it* instead. The current is large enough to blow the fuse and switch off the appliance:

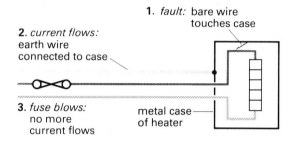

1. *fault:* bare wire touches case

2. *current flows:* earth wire connected to case

3. *fuse blows:* no more current flows

metal case of heater

Modern appliances frequently have plastic cases. Here no earth connection is necessary. The plastic case is an insulator and will not conduct electricity.

The ring main

The wires to the power sockets are laid in a loop around the house with both ends connected to the distribution box. This system has the advantage that thinner wire can be used – two sets of wire lead to each socket. Each socket is connected in parallel with the main supply as shown. The sockets are designed to give a maximum current of 13 A, but the total current taken out of a ring main must not exceed 30 A.

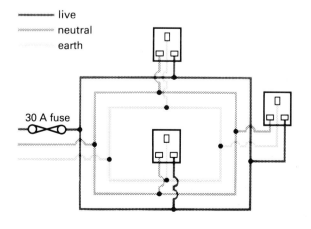

live
neutral
earth

30 A fuse

Payment for electricity

The electricity board charges for electrical energy by the kilowatt hour (kW hr), sometimes called a *unit*. This is the amount of energy used, when an appliance of power 1 kW (one thousand watts) is used for 1 hour.

$$\text{cost of electricity (in pence)} = \text{power (in kW)} \times \text{cost per unit} \times \text{time used (in hours)}$$

If the power of the appliance is given in watts, it must be divided by 1000 to get it into kilowatts.

An alternative unit of electrical energy which may be used is the megajoule (MJ). A megajoule is a million joules. When electricity is charged in megajoules:

$$\text{cost of electricity (in pence)} = \frac{\text{power (in W)} \times \text{cost per MJ} \times \text{time used (in seconds)}}{1\ 000\ 000}$$

Exercises

1. Copy out the following passage filling in the missing words from the list.

Electrical power is supplied to the * through two *. One cable is coloured * and is called the *. The other cable is coloured * and is called the *.

(neutral, cables, home, brown, blue, live)

2. Draw a diagram to show how the lights in the home are connected to the main supply.

3. Explain the purpose of earthing the metal case of an appliance.

4. Draw a diagram of a ring main.

5. If electricity costs 2p per kW hr, work out the cost of using:

(a) a 2 kW fire for 6 hours,

(b) a 3 kW kettle for ¼ hour.

(c) a 100 W light bulb for 10 hours,

(d) two 100 W light bulbs and an 800 W fire for 3 hours.

6. If electricity costs 3p per MJ, work out the cost of running:

(a) a 100 W light bulb for 3 hours,

(b) a 500 W heater for 5 hours,

(c) a 1 kW fire for 1 hour.

Questions on chapter 9

1. Four rods **a, b, c,** and **d** are charged as shown in the diagram below.

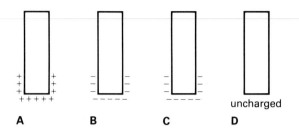

Copy down the following sentences and complete them using one of the words **attraction, repulsion,** or **no effect**.

i When **a** and **b** are brought together, there is *.
ii When **b** and **c** are brought towards each other there is *.
iii When **a** and **d** are brought towards each other there is *.
iv When **b** and **d** are brought towards each other there is *. (A.L.S.E.B.)

2. The diagram shows two metal spheres hanging by insulating threads.

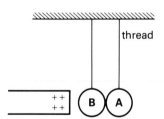

The two spheres are touching one another, and a positively charged rod has been brought up as shown.

i What charge (if any) would you expect to find on sphere **a**?
ii Explain why this sphere is charged (or not). The two spheres are now separated by pulling the insulating thread attached to **a** to the right, whilst the rod is still in position.
iii What charge (if any) would you now expect to find on sphere **b**? (W.M.E.B.)

3. Draw labelled diagrams of the cells that you would use **a** to break water up into two gases **b** to copper plate an iron nail. Explain what is happening in each case, as fully as you can.

4. a Draw a diagram to show a circuit containing two lamps in parallel, and another circuit containing the same two lamps in series.
b In which of the above circuits is the current through both the lamps the same as the current from the power supply?
c If the lamps in part **a** are identical 12 V lamps, what voltage supply will be required to light them to normal brightness when they are:
i in parallel?
ii in series?

5. The circuit below shows a 20 V supply and an ammeter, both of negligible resistance, connected to a lamp and a standard 4 Ω resistor.

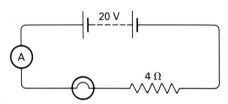

If the ammeter reads 2 A, calculate
i The resistance of the lamp,
ii The voltage across the lamp,
iii The power of the lamp. (W.J.E.C.)

6. a State Ohm's law.
b Give a large, clearly labelled diagram of the circuit you would use to show Ohm's law. What readings would you take?
c A resistance of 60 Ω has a potential difference of 12 v across it. What current passes through it?
d

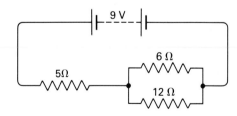

In the circuit above
i What is the combined resistance of the 6 Ω and 12 Ω resistors?
ii What is the total resistance of the circuit?
iii What is the current in the 5 Ω resistor?
iv What is the current in the 6 Ω resistor? (A.L.S.E.B.)

7. In this question you may need the formula:
$$V = I R$$
As the current through a tungsten filament lamp is increased, the potential difference across it varied in the following way:

current I in amps	potential difference V in volts
0	0
0·2	1·3
0·4	2·6
0·6	4·0
0·8	5·4
1·0	7·0
1·2	9·0
1·4	12·0

a Draw a suitable circuit for obtaining the above data.

b Plot a graph of potential difference (vertical axis) against current (horizontal axis)

c Using the graph, what is the resistance of the filament when the current I = 0·5 A?

d What happens to the temperature of the filament as the current increases?

e i What happens to the resistance of the filament as its temperature rises?
ii Explain how your answer above may be deduced from the graph.   (M.R.E.B.)

8. a i In which lead, live, neutral, or earth, is a fuse placed?
ii Give a reason for your answer.
b i Calculate the current drawn by a 2 kW appliance operating on a 250 V mains supply.
ii What value of fuse would you use for this appliance?   (N.W.R.E.B.)

9. a In the plug shown name the three pins.

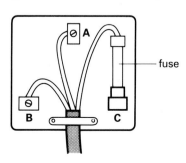

b If the international colour coding is used, name the colour on the insulation on the wire leading to
  i Pin **A**
  ii Pin **B**
  iii Pin **C**
c If the plug was connected to an electric iron, to which part of the iron would the wire from pin **A** lead?
d What would cause the fuse to melt?
(E.M.R.E.B.)

10. Calculate the cost of running a 3 kW electric fire for 2½ hours if 1 MJ (megajoule) costs 2p.
(N.W.R.E.B.)

11. An electric fire, rated as 1 kW, is connected to the 250 V mains.
  i Calculate the current taken by the fire.
  ii Which of the following cartridge fuses is most suitable for this fire, 2 A, 5 A, 10 A, or 13 A?
  iii Calculate the cost of running this fire for a week when it is switched on for 10 hours every day and electrical energy costs 2·5 pence per kilowatt-hour.   (E.A.E.B.)

12. i The diagram shows a common switching arrangement. Explain how switches **G** and **H** control the lamp. Give an example of where such a system is used.

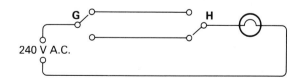

ii If the lamp in the diagram is a 60 watt lamp, what is the value of the current flowing through it when it is switched on?
iii What is the resistance of the lamp when it is working?   (S.W.E.B.)

Magnets and magnetic fields

Certain substances have the strange property called *magnetism*, and can be made into magnets.

Magnets attract iron, steel and certain other metals. Of all the elements, only three are magnetic: they are *iron, nickel* and *cobalt*. Mixtures of these metals which have been melted together to make *alloys* are also magnetic.

North and South poles. If a magnet is freely suspended, with no magnetic materials near it, it always lines up in a North-South direction. The end which points North is called the *North-seeking pole* (the N-pole), and the end which points South is called the *South-seeking pole* (the S-pole);

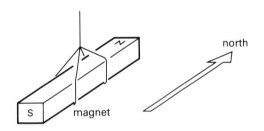

Laws of magnetism

Magnets always react to one another in the same way. This may be investigated by bringing N- and S-poles up to a magnet which is suspended by a nylon thread:

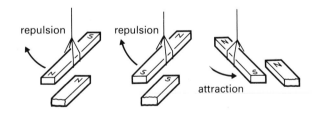

It is found that two North poles or two South poles repel each other, but North and South poles attract each other. To summarize:

Unlike poles attract and like poles repel.

If two metal bars attract one another, it is difficult to tell which one is the magnet. The only true test of a magnet is to see if it can be repelled by another magnet.

Magnetic fields

The space around a magnet where the force of attraction or repulsion is observed is called a *magnetic field*. The field is strong close to the poles of the magnet, and gets weaker further away. The magnetic field is represented by *lines of force:*

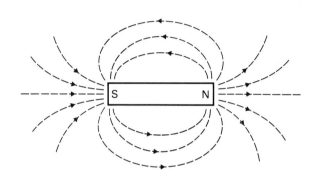

These are lines which start on an N-pole and end on an S-pole. They never touch, they never cross and they are closest together where the magnetic field is strongest.

If a small plotting compass is placed near a magnet it will always point along the direction of a line of force.

A line of force may be mapped out using a plotting compass. It is placed near to the North pole of the magnet, position **1** in the diagram below. A dot is then placed at each end of the compass needle. The compass is then moved to position **2**, so that the tail of the compass needle is next to the last dot. Another dot is then placed at the point of the needle. The compass is moved to position **3**, and a dot is again placed at the point of the needle:

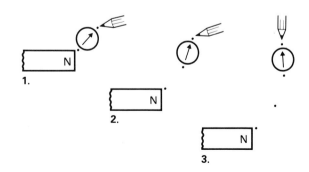

63

In this way the line can be traced round to the S-pole. The whole process may then be repeated, starting off from a different point near the N-pole. Examples of the magnetic fields round the horse-shoe magnet, and two magnets with unlike poles close together, are given below:

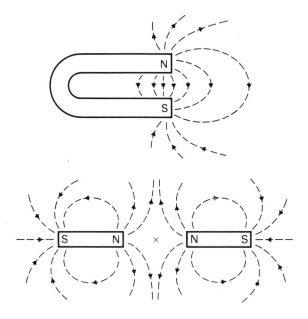

The point marked between the two N-poles is where the opposing magnetic fields cancel one another out. This point is called a *neutral point*. A compass needle placed at this point is not affected by the two magnets.

Induced magnetism

If the North pole of a magnet is brought near to an unmagnetised piece of iron, it induces magnetic poles in the iron as shown. The iron becomes a temporary magnet, as the North pole of the magnet attracts the induced South pole of the piece of iron:

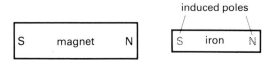

An induced magnet will also induce poles in other pieces of magnetic material. In this way a magnet can pick up two pieces of iron:

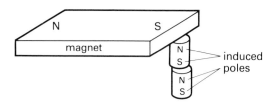

Iron filing maps. In this process, a piece of paper is placed over the magnets under investigation and iron filings are sprinkled onto it. Each filing becomes an induced magnet – as the paper is tapped they line up with the magnetic field to show the lines of force:

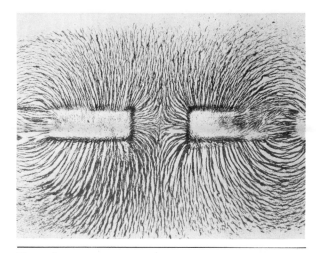

Exercises

1. Name the three magnetic elements.
2. State the laws of magnetism.
3. What is the only true test for a permanent magnet.
4. Draw the magnetic field of a bar magnet.
5. Draw the magnetic fields due to two bar magnets arranged in diagram **a** and **b** below:

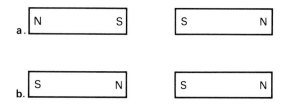

6. Which way does the needle of a plotting compass point when it is placed near to (a) the North pole of a magnet and (b) the South pole of a magnet?

The Earth's magnetism

The needle of a compass points approximately towards the North pole. This is because the Earth has a magnetic field – the compass needle points in the same direction as the Earth's magnetic lines of force. The direction in which the compass needle points is generally not exactly true North, because the Earth's magnetic poles do not coincide exactly with the Earth's geographical poles:

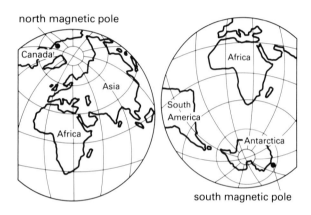

north magnetic pole

south magnetic pole

The North magnetic pole lies on an island off the North coast of Canada, and the South magnetic pole lies in the sea off the coast of the Antarctic continent.

Magnetic declination. The angle between true North and the direction in which the compass needle points is called the *magnetic declination*. The angle of declination varies over the world – the angle is particularly large in regions near the magnetic pole:

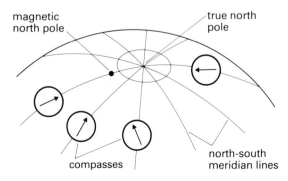

magnetic north pole

true north pole

compasses

north-south meridian lines

In the United Kingdom the magnetic compass points in a direction which is about 7° West of true North – the *magnetic declination* is 7° West of North:

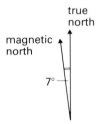

true north

magnetic north

7°

Compasses

The magnetic compass is still the most reliable method of finding direction:

The "Silva" compass. This type of compass should be carried by all mountaineers and hill walkers:

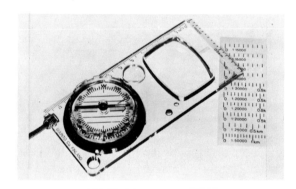

The compass needle is mounted in a clear plastic case which is filled with alcohol. The alcohol steadies the movement of the needle – friction between it and the liquid stops it from swinging, so that it settles down quickly.

The mariner's compass. This is still widely used on small ships. One is often carried even on large ships that use the more modern gyroscopic compass, in case the modern one breaks down.

The mariner's compass differs from an ordinary one. The compass needle is mounted into the underside of a circular card. The top of the card is marked with the points of the compass. A reference point is marked on the compass case. This reference mark points directly along the line of the ship. The reading on the compass card next to the reference point gives the direction in which the ship is heading, as shown in the diagram at the top of the next column.

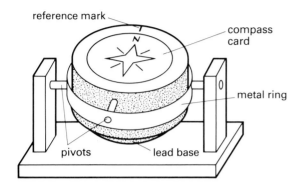

64

The compass card is mounted in a non-magnetic bowl with a lead base. The bowl is fixed by two pivots to a metal ring, which is in turn fixed by two other pivots at right angles to the first. This is called gimbal mounting. Because of this arrangement, no matter how the ship pitches and rolls, the bowl and compass card stay horizontal.

The Earth as a magnet

The cause of the Earth's magnetic field is not understood. However, the magnetic field can be described by imagining that there is a giant bar magnet at the centre of the Earth. The imaginary magnet is about 4000 km long with its South pole pointing towards the Earth's North geographical pole. The Earth's magnetic field is shown in the diagram below:

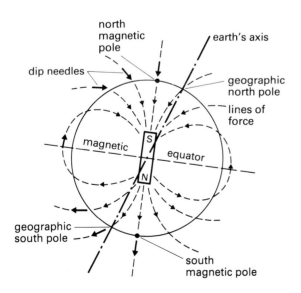

If a compass needle is freely suspended so that it can point in any direction, it points in the direction of these lines of force. In the United Kingdom the needle dips down at a steep angle of about 67° from the horizontal. This angle is called the *angle of dip*. The angle of dip at the "magnetic equator" is 0° because the lines of force at the magnetic equator are parallel to the Earth's surface. The angle of dip at the magnetic pole is 90° because the lines of force enter the earth vertically.

The angle of dip is measured with a dip circle similar to the one shown in the photograph. It consists of a compass needle pivoted in a vertical plane:

Exercises

1 Why does a compass needle generally not point in the direction of true North?
2. What is meant by the angle of declination and the angle of dip?
3. What is the purpose of mounting a compass needle in a case that is filled with liquid?
4. How does a Mariner's compass differ from an ordinary one?
5. Draw a diagram to show the Earth's magnetic field. Mark on the diagram the places where the angle of dip is 0° and 90°.

Electromagnetism

If a compass is placed near a wire carrying an electric current, the needle is deflected – an electric current has a magnetic field associated with it. The field can be mapped out using a plotting compass. Its shape depends on the shape of the wire. The fields round three shapes of wire are shown below.

Field round a straight wire. The lines of force are circles around the wire. Their direction depends on which way the current flows:

The direction of the lines of force can be worked out from the *right hand grip rule.* Imagine the wire gripped in the right hand with the thumb pointing along the wire in the direction of the conventional current, then the direction of the fingers will give the direction of the lines of force:

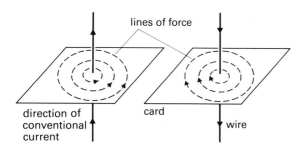

Field due to a narrow coil. The direction of the circular lines of force near to the wire are worked out from the right hand grip rule.

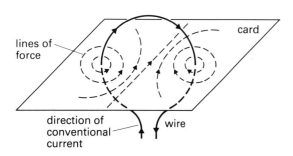

Field due to a long coil or solenoid. The magnetic field of a solenoid is made up of the fields of a series of narrow coils. They add together to make a strong field like that of a bar magnet. One end of the solenoid is an N-pole and the other end an S-pole:

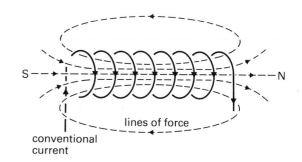

Which end is which, can be worked out by looking at the end of the coil. If the current is going round in a clockwise direction that end is a South pole, as in diagram **1** below. If the current is going round in an anticlockwise direction that end is a North pole, as in diagram **2**:

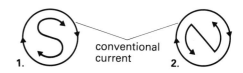

65

Electromagnets

If a bar of pure iron is placed inside a solenoid, it becomes strongly magnetised when the current is flowing. When the current is switched off the iron loses its magnetism. This device is called an *electromagnet*. Electromagnets are used in many things such as electric bells, electric buzzers, and magnetic switches called relays.

The electric bell and buzzer

The electric bell uses an electromagnet which switches on and off automatically, causing a hammer to repeatedly strike a gong. As soon as the current is switched on, a current flows across the contacts, and round the coils of the electromagnet:

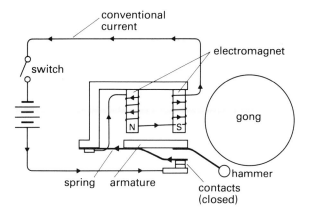

The electromagnet attracts a piece of iron called the *armature*. This makes the hammer strike the gong. But as it does so, the contacts separate – this stops the current flowing, and switches off the electromagnet:

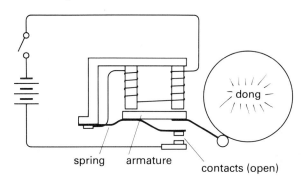

The armature and hammer now spring back closing the contacts again, as shown in the first diagram. This allows the current to flow again so that the whole cycle starts again, and continues until the

bell is switched off. The buzzer works in exactly the same way except that there is no gong – the vibration of the armature causes the sound.

The electric relay

This is a magnetic switch which uses a small current in a *primary circuit* to control a larger current in a *secondary circuit*. When the primary circuit is switched on, the electromagnet attracts the armature so that its top end rises, pushing the contacts together. This allows a current to flow in the secondary circuit:

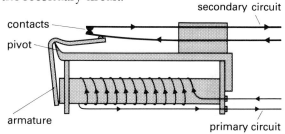
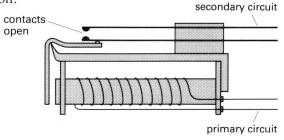

As soon as the primary circuit is switched off, the spring of the lower contact pushes the top of the armature down again. The contacts are now open and the current in the secondary circuit is switched off.

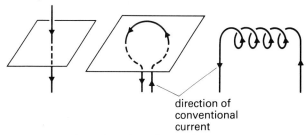

Exercises

1. Copy the diagrams below, draw in the lines of force, and mark their direction.

direction of conventional current

2. Draw a diagram of an electric buzzer and explain how it works.
3. With the aid of a labelled diagram explain how an electric relay works.

Theory of magnetism

If a magnet is broken in two, this does not make a separate North pole and South pole but two similar magnets. No matter how many times the magnet is broken smaller magnets are always obtained:

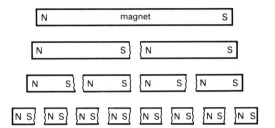

Experiments suggest that in magnetic materials the molecules themselves are small magnets, so that it is impossible to get a separate North or South pole. The theory of magnetism is based on this idea of "molecular magnets".

The domain theory of magnetism

This theory suggests that in magnetic materials the molecular magnets line up with their North and South poles together to make highly magnetised regions called *domains*. The diagram below shows a piece of unmagnetised material such as a piece of iron. The molecular magnets are drawn as arrows, with the arrow head at the North pole. The molecular magnets in each domain are pointing in different directions. Because of this the Norths and Souths of the molecular magnets cancel one another out:

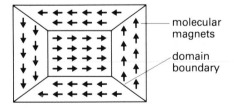

If this unmagnetised material is placed in a weak magnetic field the molecular magnets in some of the domains turn to point in the direction of the magnetic field. The poles of some of the molecular magnets at either end of the material are no longer cancelled out and so it becomes a weak magnet, as shown in the next diagram:

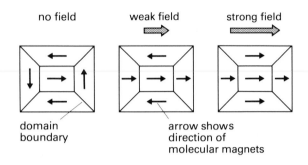

As the applied magnetic field is increased, the molecular magnets in more domains turn to point in the direction of the field, until all the domains are pointing in the same direction. The material can be magnetised no further and is said to be *magnetically saturated*.

Magnetic properties of different materials

If a permanent magnet is brought near to iron and steel bars it induces magnetic poles in them. These will attract iron filings:

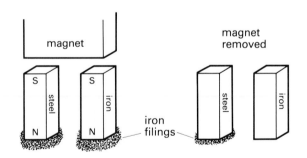

When the permanent magnet is removed, the iron filings fall off the iron. Some iron filings still stick to the steel – it has become a weak magnet. For the iron, the field of the permanent magnet can easily line up its molecular magnets, making it a temporary magnet. When the permanent magnet is removed the molecular magnets in the iron return to their old positions and the iron loses its magnetism. A material like iron, which is magnetised easily and loses its magnetism easily, is said to be *magnetically soft*.

In the case of steel, the molecular magnets are more difficult to line up but once they have been lined up they stay in that position. Materials like steel which retain their magnetism are said to be magnetically hard. Permanent magnets are made from *magnetically hard* materials.

Methods of making magnets

There are three methods of making magnets. The best method is described first.

1. Using a coil. The strong magnetic field generated by the long coil lines up all the molecular magnets in the steel bar:

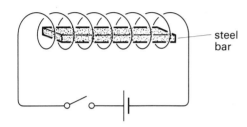

steel bar

2. Stroking with a magnet. The steel bar is stroked with a permanent magnet many times in the same direction. The stroking gradually lines up the molecular magnets in the steel. The end of the steel bar at which the stroking starts becomes the same as the stroking pole:

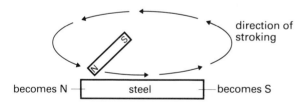

direction of stroking

becomes N — steel — becomes S

3. Tapping. If a steel bar is placed pointing North-South and tapped gently for a long time, it becomes a weak magnet. The tapping jogs the molecular magnets so that some line up with the Earth's magnetic field.

Methods of demagnetising

Again, there are three methods of demagnetising and the first one is the best.

1. Heating. If a magnet is heated in a bunsen it loses its magnetism. As the temperature rises the vibrations of the molecular magnets become so great that their order is destroyed, as in diagram **1** at the top of the next column.

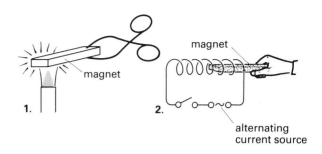

magnet

1.

2.

alternating current source

2. Using and A.C. coil. An alternating current (A.C.) which changes its direction 100 times a second is passed through the coil. The magnetic field of the coil changes at the same rate. As the magnet is slowly withdrawn from the coil the changing magnetic field destroys the order of the molecular magnets, as shown in diagram **2**.

3. Hammering. If a magnet is hammered, the blows destroy the order of the molecular magnets and it loses it magnetism.

Exercises

1. Use the domain theory of magnetism to explain how the magnetic field of a coil magnetises a steel bar.

2. How does the domain theory explain magnetic saturation?

3. Copy out the following sentences filling in the missing words from the lists.

A magnetically * material is one which is magnetised easily and * its magnetism *.
(loses, soft, easily)

A magnetically * material is one which is difficult to * but * its magnetism.
(retains, magnetise, hard)

4. There are three ways of making magnets. Describe each one in a sentence.

5. There are three ways of demagnetising. Describe each one in a sentence.

Applications of electromagnetism

The magnetic force produced by an electric current is used in many ways. This section describes three of those uses.

The moving iron meter

The moving iron meter measures electric current. It uses the force of repulsion between two iron bars, which are both magnetised by the current flowing round a coil. One iron bar is fixed whilst the other is attached to the same rod as the pointer. The pivots for the rod are omitted from the diagram for clarity. The tension in the fine coil spring keeps the iron bars together when no current is flowing:

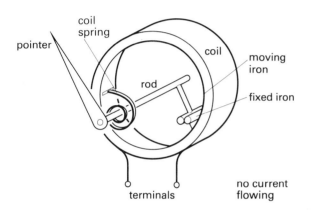

When the current that is being measured is passed round the coil it magnetises both iron bars the same way, with like poles next to one another. The strength of these poles, and hence the force of repulsion between them, depends on the strength of the current. The moving iron is repelled from the fixed one until the force of repulsion is balanced by the tension of the coil spring. The position of the pointer is a measure of the current:

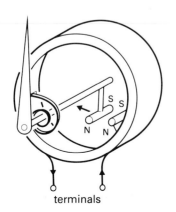

The moving iron meter has the advantage that it can be used to measure both alternating currents, which continually change their direction, and direct currents which flow in one direction only. The reason for this is that no matter which way the current flows round the coil, the irons are always magnetised with like poles together. The moving iron meter has the disadvantage that the scale does not increase steadily. It is squashed up at each end:

The telephone

The microphone part of a telephone uses a variable resistor, whose resistance changes as sound waves hit it. It controls the varying current that is sent to the earpiece. The earpiece of a telephone uses an electromagnet to make a thin sheet of iron vibrate and produce sound.

67

Microphone. The variable resistor is made of two carbon blocks that are separated by carbon granules. In the microphone one of the blocks is attached to a diaphragm. The sound waves strike against a thin metal sheet called a *diaphragm,* making it vibrate. The diaphragm moves the carbon block attached to it. As the carbon blocks are moved closer the granules are compressed together. More granules are now in contact with one another so that there are more paths for the electricity to flow through – the resistance becomes lower. If the carbon blocks are moved apart the granules are loosened and the resistance is higher:

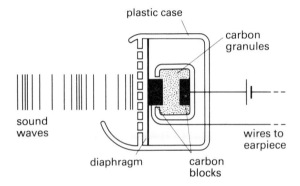

The resistance therefore varies with the compressions and rarefactions of the sound wave making the current supplied by the battery vary in the same way.

Earpiece. The varying current from the microphone passes through the electromagnet of the earpiece. The varying current alters the strength of the electromagnet. The electromagnet attracts the iron diaphragm by different amounts, depending on its strength. The diaphragm therefore vibrates in sympathy with the changing current, and reproduces the sound which entered the microphone:

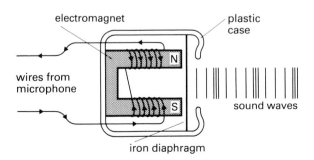

The moving coil loudspeaker

The moving coil loudspeaker works in a similar way to the telephone earpiece. The loudspeaker is designed to give a louder and much more faithful reproduction of the sound. In this case, the coil is inside the strong magnetic field from a permanent magnet. Attached to the end of the coil is a large, light paper cone. When a varying current is passed through the coil it produces a varying magnetic field, which pushes the coil away from the permanent magnet by an amount depending on the field strength. In this way, the paper cone vibrates according to the varying current, and reproduces the sound heard in the microphone:

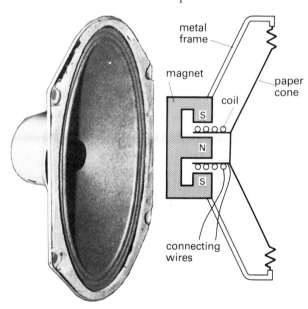

Exercises
1. Draw a diagram of a moving iron meter and explain how it works.
2. Explain with the aid of diagrams
(a) how the telephone microphone works and
(b) how the telephone earpiece works.
3. Draw a diagram of a moving coil loudspeaker and explain how it works.

Questions on chapter 10

1. Three metal bars are labelled **A**, **B**, and **C**. Each end of **A** attracts each end of **B**. One end of **B** attracts one end of **C**, but it repels the other end of **C**.

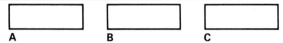

A   **B**   **C**

i What can be deduced about the bars **A**, **B**, and **C**?

ii Give reasons for these deductions.

iii Describe a further test which could be tried on all three bars to confirm the deductions. Give the result of the test in each case.   (Y.R.E.B.)

2. The diagram shows the Earth and its magnetic field.

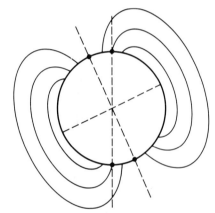

Copy the diagram into your book.

  i Mark with arrows the directions of the lines of magnetic force.

  ii Mark with **N** the north magnetic pole.

  iii What would you use to find the direction of the Earth's magnetic field?

  iv Mark with **D** one place on the Earth's surface where the angle of dip is zero.   (E.M.R.E.B.)

3. The diagram shows a straight vertical wire passing through a horizontal sheet of cardboard. A current of about 10 A passes along the wire in the direction indicated by the arrow.

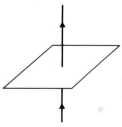

i Copy the diagram and sketch the magnetic field near the wire. State any rule which could be used to find the direction of the lines of force in this field and the result of applying the rule in this case.

ii The shape of the lines of force could be determined by the use of iron filings. Explain how iron filings would work and why copper turnings would not.

iii Draw a labelled diagram of a magnetic compass.

iv Explain how such a compass could be used to determine the shape and direction of the magnetic field near the wire shown. (W.Y. & L.E.B.)

4. Look at the following diagram:

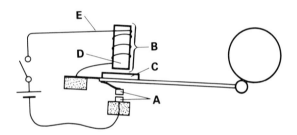

i Name the parts **A** and **B**.

ii Name the substances used in constructing the parts **C**, **D**, and **E**.

iii Describe the sequence of events following the closing of the switch and so explain how the bell works.   (Y.R.E.B.)

5.

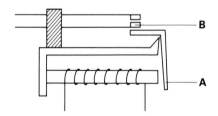

i What is the name of the device shown in the diagram?

ii Name the parts labelled **A** and **B**.

iii Name a suitable material for **A**.
iv Explain briefly how the device works.
v Suggest one application of the device.
(W.M.E.B.)

6. i State two differences in the magnetic properties of iron and steel.
ii Describe two non-electrical methods of making a bar of steel into a magnet. (W.J.E.C.)

7.

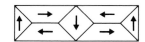

The diagram shows a simplified view of a piece of unmagnetised steel. The arrows show the direction of magnetisation of each domain. Copy the diagram into your book, and label it "unmagnetised".
i Draw a similar diagram to show the domains as they would be when the steel is partly magnetised. Label it "partly magnetised".
ii Draw another diagram to show the domains as they would be when the steel is fully magnetised. Label it "fully magnetised". (Y.R.E.B.)

8.

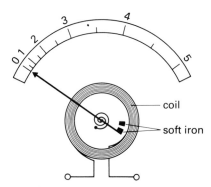

The diagram shows a simple electrical current measuring instrument (ammeter).
i What type of ammeter is shown?
ii Say whether it is capable of measuring alternating current only, direct current only, or both.
iii Explain how the instrument shown works.
(N.W.R.E.B.)

9. i What is the purpose of a microphone?
ii Explain with the aid of diagrams, the working of a simple carbon microphone.

iii The diagram below shows a type of home-made microphone. Explain fully how an output is produced when someone speaks into the paper cone.

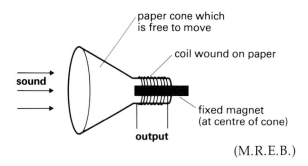

(M.R.E.B.)

10. An incomplete diagram of a moving coil loudspeaker is drawn below.

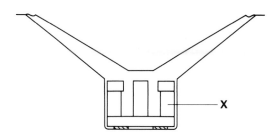

Copy the diagram into your book and complete it. What is the part labelled **X**? (E.A.E.B.)

11. Copy the diagram of the telephone ear piece into your book and complete it.

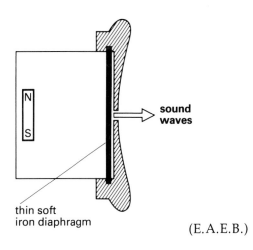

(E.A.E.B.)

Electric motors

There is a force on a current carrying wire in a magnetic field. The electric motor makes a large force by placing many such wires in a magnetic field. The actual way it works depends on the way that a single wire acts.

Force on a single wire

In the apparatus shown a magnetic field is provided by a horse-shoe magnet. The wire which passes between the poles of the magnet, is mounted on conducting pivots so that it is free to swing:

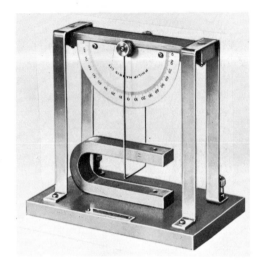

When an electric current is passed through the wire there is a force on it causing it to swing one way or the other. The direction of the force depends on the direction of the magnetic field:

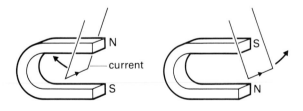

The direction of the force also depends on the direction of the current:

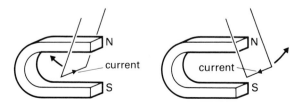

Fleming's left hand rule. This rule summarises what is happening to the wire. It says that the directions of the force, current and magnetic field are all *at right angles* to one another. The actual directions can be worked out by pointing the forefinger, centre finger and thumb of the left hand at right angles to one another:

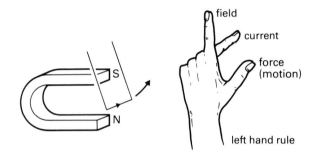

If the Forefinger is pointed in the direction of the Field, and the Centre finger in the direction of the Current, the thuMb points in the direction of the Motion of the wire. The motion of the wire is in the same direction as the force. Summarising all this:

Forefinger   : Field
Centre finger : Current
thuMb   : Motion

Strength of the force on the wire. It is found that when the current through the wire is increased, or two magnets are used instead of one, then the wire swings further. The strength of the force increases with both the strength of the current and the strength of the magnetic field.

The electric motor

The simple electric motor consists of a coil of wire mounted on a rod, so that it is free to rotate in a magnetic field.

The magnetic field is provided by permanent magnets. A current is passed into and out of the coil through *brushes* made from carbon or copper, which press against a *split ring commutator,* made from two semicircles of copper fixed onto an insulating cylinder. Each end of the coil is joined to one of the split rings:

68

In some motor cycles, the same fitting acts both as a starter motor and as a generator. (The next section explains how.)

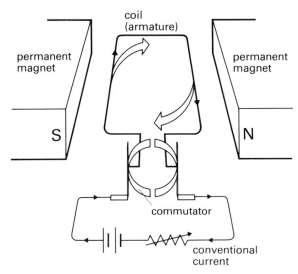

coil (armature)

permanent magnet

permanent magnet

S

N

commutator

conventional current

In the arrangement shown in the diagram above, the current enters by the left hand brush, goes round the coil, and leaves by the right hand brush. The magnetic force only acts on the parts of the coil that go "into the paper" and "out of the paper". Fleming's left hand rule suggests that the force on the wire going "into the paper" (next to the S-pole) is upwards and that coming out of the paper (next to the N-pole) is downwards. This turns the coil in a clockwise direction.

As the coil turns over, the connections to the commutator are reversed – the current flows through the coil in the opposite direction – but it *still* flows "into the paper" by the S-pole and "out of the paper" by the N-pole.

The force on the coil is the same as it was before – upwards on the left hand side and downwards on the right hand side. *No matter what the position of the coil,* the force on it is *always* clockwise. The motor will continue to rotate until the current is switched off.

Practical motors are made with split rings divided into many parts, joined to many coils. This makes the motor smoother running and more powerful.

Factors affecting the power of an electric motor.
The force which turns the coil of the simple electric motor can be increased in three ways:
1. By increasing the strength of the magnetic field. This is usually done by replacing the permanent magnets with electromagnets.
2. By increasing the current flowing. This can be done with the variable resistance in the circuit.
3. By winding the armature with more coils of wire. The force is the same on each turn of wire. Therefore, the total force is increased by increasing the number of coils.

Exercises
1. What factors affect the strength of the force on a current carrying wire in a magnetic field?
2. Name and explain the rule which gives the direction of the force on a current carrying wire in a magnetic field.
3. Draw a diagram of a simple electric motor and explain how it works.
4. Explain how the power of an electric motor can be increased.

Electric generators

The *dynamo effect* used to produce electricity is the opposite of the "motor effect" described in the previous section. The dynamo effect can be demonstrated by connecting the "force on a conductor apparatus" to a sensitive current meter as shown:

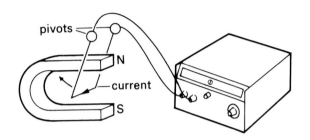

It is found that when the conductor is swung to and fro through the magnetic field the meter registers a current, first in one direction and then the other. A current is *induced* in the wire. The effect is called *electromagnetic induction*.

Size of the induced current. It is found that the induced current is larger when:

1. the wire moves faster through the magnetic field;

2. the wire moves through a stronger field that has been made by using two magnets;

3. a double loop of wire is passed through the magnetic field as shown below. The currents induced in each length of the wire cutting the magnetic field add together to make twice the current.

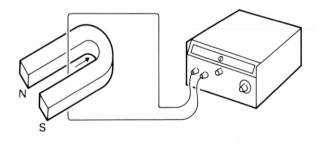

4. the wire cuts across the magnetic field as shown in diagram **1** at the top of the next column – rather than moving parallel to it as in diagram **2**.

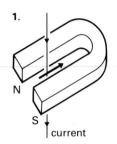

These results are summarised by *Faraday's law of electromagnetic induction* which states that the size of the induced current is proportional to the rate at which each length of wire cuts through magnetic field.

The direction of the induced current. This is best demonstrated by means of a magnet and a coil. When the N-pole of a magnet is pushed into the coil making the wires cut through the magnetic field an induced current flows so that the end of the coil next to the magnet becomes an N-pole. The force of repulsion tries to stop the magnet being pushed in:

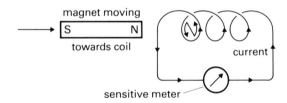

When the magnet is pulled out the current flows in the opposite direction so that the end of the coil becomes an S-pole. The force of attraction tries to stop the magnet being pulled out:

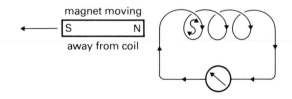

A force always has to be used to move the magnet, and it is this force which drives the current around the circuit.

These experiments are summarised by *Lenz's law* which states that the induced current flows in a direction that tries to stop the change in magnetic field which produces the current.

69

The dynamo

The dynamo is a motor being used in reverse. It consists of a coil of wire that is forced to cut through a magnetic field. As the wire cuts through the magnetic field, a current is induced in it.

The only parts of the coil that cut through the magnetic field as the coil is rotated, are those parts which are parallel to the faces of the magnet. In diagram **1** the wires are sweeping through the magnetic field at the maximum rate, and the induced current is a maximum:

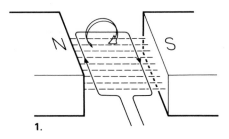

1.

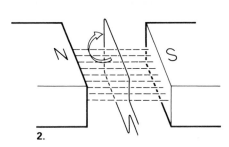

2.

As the wires sweep round to the top and bottom positions as in diagram **2** they are moving parallel to the magnetic field and not cutting through it – there is no induced current. Then, as the coil turns over, the wire which was sweeping up through the magnetic field is now sweeping down through it: the induced current flows in the opposite direction. The graph below shows how the current in the coil varies with the position of the coil. This is an *alternating current*:

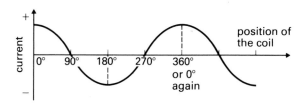

An alternating current (AC) dynamo is made by connecting the coil to the load (heater, motor etc.) by means of two slip rings. The same side of the coil is then always connected to the same side of the load. Therefore the current in the load changes in the same way as the current in the coil:

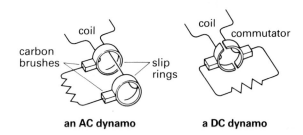

an AC dynamo   **a DC dynamo**

In a direct current, or DC dynamo the coil is connected to the load by a split ring commutator. This reverses the connections to the load every half turn of the coil so that the current always flows through the coil in the same direction, as shown in the graph below:

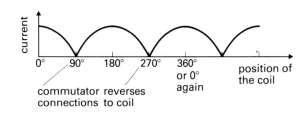

Exercises

1. Copy the diagram into your book and mark the direction of the induced current in the coil. Name and state the law which enabled you to work out the direction of the current.

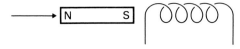

2. What factors affect the size of the induced current in a single wire that is moved through a magnetic field?
3. Draw a diagram of an A.C. dynamo. Label the slip rings, the carbon brushes, the coil and the load.
4. Draw a diagram of a D.C dynamo. Label the split ring commutator, the carbon brushes, the armature and the load.

Transformers

The previous section explains how moving a permanent magnet towards a coil sets up an induced current in the coil. An induced current is still set up when the permanent magnet is replaced by an electromagnet:

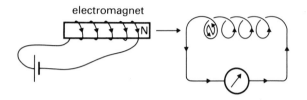

The changing magnetic field can also be provided by fixing the coil next to the electromagnet and switching it on and off:

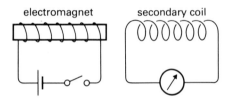

An induced current is produced both when the electromagnet is switched on, and when it is switched off.

If an alternating current which is continually changing its direction is passed through the coil of the electromagnet, a magnetic field is produced. The continually changing magnetic field induces a continually changing current in the coil. This effect is very important – it is used in the *transformer* to convert a high potential difference into a low potential difference and vice versa.

The transformer
The transformer shown at the top of the next column consists of two coils of wire wound on a soft iron core. This is made up of laminas (thin sheets) of soft iron separated by layers of varnish. The purpose of the soft iron core is to concentrate the magnetic fields produced.

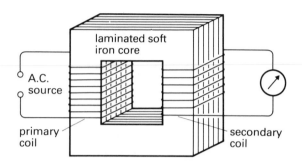

An alternating current is passed through the *primary coil* which acts as an electromagnet, and generates a magnetic field which is continually changing. The changing magnetic field passes round the soft iron core and through the *secondary coil* where the changing field induces a secondary current.

The potential differences (P.D.'s) across the primary and secondary coils are found from the equation:

$$\frac{\text{P.D. across secondary}}{\text{P.D. across primary}} = \frac{\text{Number of turns on secondary}}{\text{Number of turns on primary}}$$

$$\frac{V_s}{V_p} = \frac{N_s}{N_p}$$

Transformers are divided into two types as shown in the diagram. A step-up transformer has more turns of wire on the secondary coil than on the primary coil and *increases* the potential difference. A step-down transformer has less turns of wire on the secondary coil than on the primary coil and *decreases* the potential difference:

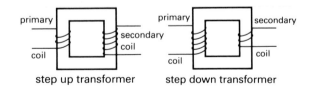

Transformers are made in a complete range of sizes from very small to very large. The world's first transformer was made by Michael Faraday in 1831. The photograph at the top of the next column shows a modern power transformer.

70

Energy lost in a transformer. Energy is lost in three ways:

1. Resistance heating of the wire of the coils. This is reduced by making the coils of thicker wire so that their resistance is less.

2. The changing magnetic field not only induces a current in the secondary coil, but also in the iron core. The currents induced in the core are called eddy currents and they heat the core up. Eddy currents are reduced by *laminating* the core – making it of many thin sheets of iron separated by layers of varnish to stop the eddy currents flowing.

3. Energy is lost in continually changing the direction of the molecular magnets in the iron core.

If the transformer is well designed, the energy losses are small, so:

power put into _ power got out of
primary coil ‾ secondary

In other words:

P.D × current _ P.D. × current
in primary ‾ in secondary

$$V_p I_p = V_s I_s$$

This means that the potential difference multiplied by the current is the same in each coil. If there is a high potential difference and a low current in the primary then there is a low potential difference and high current in the secondary, or vice versa.

Transmission of power

The cables carrying electricity across the country are miles long, and power is lost because the electric current heats the cable up. The power lost in heating the cables is given by:

power *lost* = resistance × (current)²
= RI^2

the power *carried* by the cable is given by:

power   potential
carried =   difference × current

= VI

A step-up transformer such as that shown is used to increase the potential difference V and reduce the current I for transmission along the cable. The power carried, which is VI stays the same. However the power lost RI^2 is much smaller because I is smaller. A step down transformer is used at the other end of the cable to reduce the potential difference again.

The diagram below shows how power is transmitted from the power station to industry and houses:

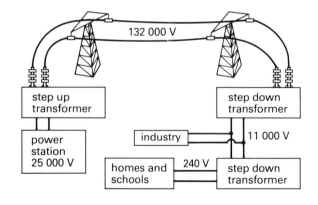

Exercises

1. Draw a diagram of a step down transformer. Label the primary coil, the secondary coil and the laminated soft iron core. Explain how the transformer works.
2. In what three ways is power lost in a transformer.
3. Explain the advantage of transmitting power at a high potential difference.

The moving coil meter

The moving coil meter works on the same principle as the electric motor. It consists of a coil of wire which is pivoted so that it can rotate in a magnetic field. At each end of the coil there is a coil spring through which the current to be measured enters and leaves:

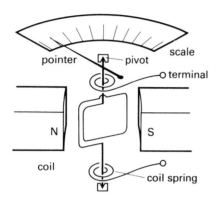

When a current passes through the coil there is a force on it, which makes it turn. The coil turns until the magnetic force on it is balanced by the force due to the tension of the coil springs. The position of the coil is then a measure of the current. A pointer is fixed to the coil so that the current can be read off from a scale.

The scale of the moving coil meter is made uniform or *linear* by fixing a cylinder of soft iron inside the coil, as shown in the plan view. This has the effect of making the magnetic field uniformly strong in the gap through which the coil moves:

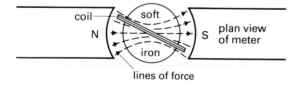

The direction in which the coil turns depends on the direction of the current, so the meter will not work with alternating currents which change direction rapidly.

The photograph at the top of the next column shows a demonstration moving coil meter. The coil spring and the pointer are clearly visible. The permanent magnet and the soft iron core are behind the scale:

The basic meter, which only takes a very small current, is called a *galvanometer*. The meter has to be specially converted to make it into an ammeter or a voltmeter.

Conversion of a galvanometer into an ammeter or a voltmeter

This conversion is carried out by connecting resistors of different values either in parallel or in series with the meter. The resistors may be inside the meter case or connected by means of plug in attachments as in the photograph:

This galvanometer takes 100 μA (0.0001 A) to make the pointer move right across the scale and has a resistance of 1000 Ω.

71

Conversion to an ammeter. A low resistance called a *shunt* is connected in parallel with the meter. Most of the current to be measured takes the easy path through the shunt. Only a very small part of the current, which is just enough to work the meter, actually flows through the coil of the meter. If the meter is to read to 1 A, the low resistance must be chosen so that 0.9999 A flows round the shunt. Then only 0.0001 A flows through the meter:

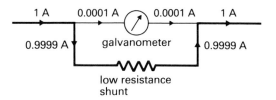

If the meter is required to read a higher current a lower resistance shunt must be used. The lower resistance is needed so that an even larger part of the current to be measured flows through the shunt.

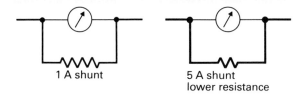

Conversion to a voltmeter. A high resistance called a *multiplier* is connected in series with the meter. In order to measure the potential difference across a bulb, for example, the meter must be connected as shown:

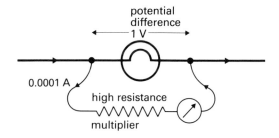

In this case the meter is required to read a maximum potential difference of 1 V. The value of the high resistance multiplier is chosen so that, when there is a potential difference of 1 V across the meter and multiplier, only the 100 μA (0.0001 A) needed to operate the meter flows through it.

The value of the resistance of the multiplier is higher, when the meter is required to measure higher potential differences. The higher resistance is necessary to prevent the higher potential difference driving more than 100 μA through the meter.

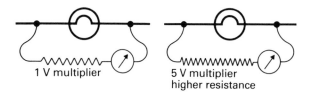

The multimeter

This is a galvanometer which has several different multipliers and shunts to provide different ranges of potential difference and current. The various resistors are connected in turn to the meter by the rotary switch on the front of the meter case:

Exercises

1. Draw a diagram of a moving coil meter and explain how it works.
2. Explain why a moving coil meter cannot be used to measure alternating currents.
3. Explain how a moving coil galvanometer can be converted into (a) an ammeter and (b) a voltmeter.

The ignition system of a car

The petrol in the cylinders of the car engine is ignited by an electric spark. A typical spark plug is shown below:

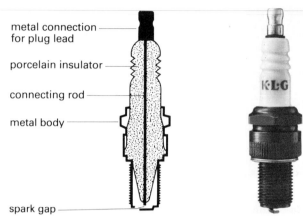

metal connection for plug lead

porcelain insulator

connecting rod

metal body

spark gap

The spark jumps across a gap of about 0.6 mm between the central electrode and the body of the spark plug. The central electrode is insulated from the body by a porcelain sleeve. A potential difference of about 10 000 V is used to drive the spark across the gap. The 10 000 V is produced by a special type of transformer called an *induction coil* which is shown at the bottom left of the photo:

Can you recognise any other of the electrical parts mentioned in the text?

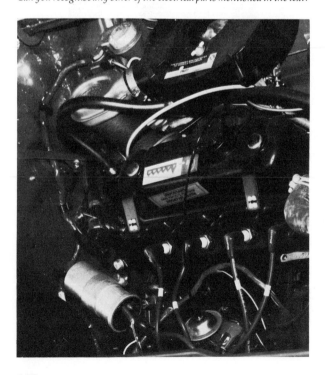

The principle of the ignition system is shown in the diagram below. The induction coil has a core made of soft iron wires. The primary coil consists of a few turns of thick wire. The secondary coil has many thousands of turns of thin wire:

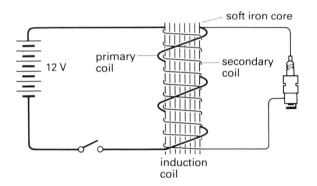

soft iron core

12 V

primary coil

secondary coil

induction coil

When the switch in the primary circuit is opened, the magnetic field reduces rapidly. This induces the high potential difference across the secondary coil which produces the spark. When the switch is closed again the magnetic field does not build up as quickly – the induced potential difference across the secondary coil is then not high enough to cause a spark. Therefore there is only a spark when the switch is opened.

In a car the switch is driven from the engine. A square piece of metal called a *cam,* rotates and forces the contacts apart just as the piston reaches the top of the compression stroke, and the petrol is ready to be exploded:

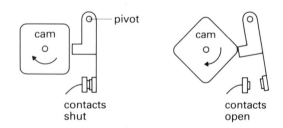

pivot

cam

contacts shut

cam

contacts open

A car engine usually has four cylinders and four spark plugs. The correct spark plug must be connected to the 10 000 V side of the induction coil at the right time. The *distributor* does this. Its main part is an arm, called a *rotor arm* which rotates making contact with each spark plug in turn. The rotor arm is mounted on the same shaft as the contact breaker cam:

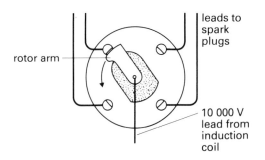

The complete ignition circuit is shown below. As the contact breaker cam opens the contacts, the rotor arm connects the secondary winding of the induction coil to one of the plugs. The current in the primary coil switches off. The rapidly reducing magnetic field induces the high potential difference across the secondary coil which produces the spark. A *capacitor* in the primary circuit helps to prevent sparking across the contacts of the contact breaker:

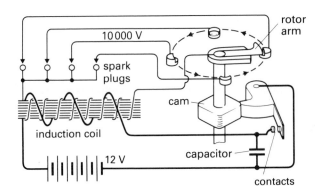

The laboratory induction coil

This is used to step up the potential difference of a battery to 20 000 V and more. The changing magnetic field is produced by switching the current in the primary coil on and off in the same way as in the electric bell, as shown in the diagram at the top of the next column.

When the primary current is switched on, it flows across the contacts and around the primary coil. This magnetises the soft iron wires, which attract

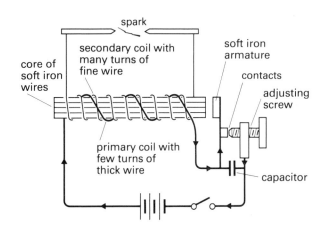

the armature, and pull the contacts apart switching off the current. The iron wires lose their magnetism, and the armature springs back closing the contacts again and switching on the current.

In this way a current and a magnetic field are produced which are continually switching on and off. The changing magnetic field passes through the many thousands of turns of the secondary coil inducing a high potential difference, which can produce a continuous spark.

Exercises

1. Draw a labelled diagram of a spark plug.
2. Draw a labelled diagram of the car ignition system and explain how it works.
3. Explain with the aid of a labelled diagram how the laboratory induction coil works.

The car electrical system

The last section described the ignition system of the car. This section describes the other electrical systems. They use many of the principles and devices that have been described in previous sections.

On a car there is only one wire from the battery to each piece of electrical equipment. This is possible because the car body and chassis is made of steel. Steel is an electrical conductor – the return path to the battery is through the chassis.

The charging system

In a modern car the electrical energy is produced by a type of dynamo called an *alternator*:

The alternator, which is driven by the engine, produces an alternating current. This is converted electronically into a direct current, which is used to charge the battery. The battery acts as a reservoir or store of electricity – all the electrical equipment is operated from the battery. The current generated by the alternator is automatically controlled so that less current is produced when the battery is well charged:

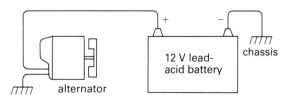

The starter motor

The starter motor is a powerful electric motor which takes a current of over 30 amps. The wire from the battery to the motor has to be very thick in order to carry this large current. The current for the starter motor is switched on by means of a relay. The first turn of the ignition key makes current available in the ignition system, a second turn will operate the relay:

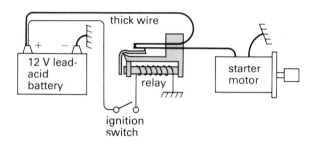

The fuse box

All the other electrical equipment is connected to the battery through a fuse box:

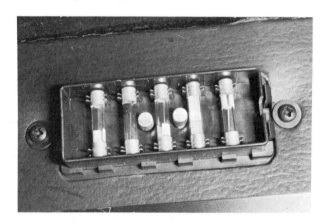

If a positive wire should come into contact with the metal chassis, the large current that flows, will blow the fuse. If there was no fuse this large current would flow until the battery was flat. The diagram shows how the electrical equipment is connected:

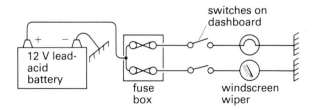

Direction indicator lights

The diagram shows how the direction indicator lights are connected. The flasher unit is the device which makes the lights flash on and off. When the two way switch, which is on the steering column, is in the up position as shown, the left hand lights flash. When the two way switch is in the down position the right hand lights flash:

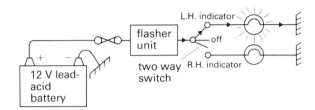

A flasher unit contains a bimetallic strip. When the contacts are closed a current flows through the heater and lights the bulb:

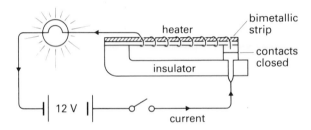

As the bimetallic strip heats up, it bends upwards separating the contacts. The current is switched off and the bimetallic strip cools. As it does so, it bends back again closing the contacts and switching on the current:

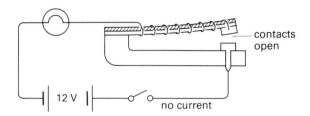

In this way, by bending up and down the bimetallic strip continually switches the indicator lights on and off.

Exercises

1. Explain how the car battery is charged up.
2. Explain with the aid of a diagram how the starter motor is switched on.
3. What is the purpose of having a fuse in the wires leading to the electrical equipment?
4. Explain with the aid of a diagram how the flasher unit works.

General view of the Austin Morris Marina engine.

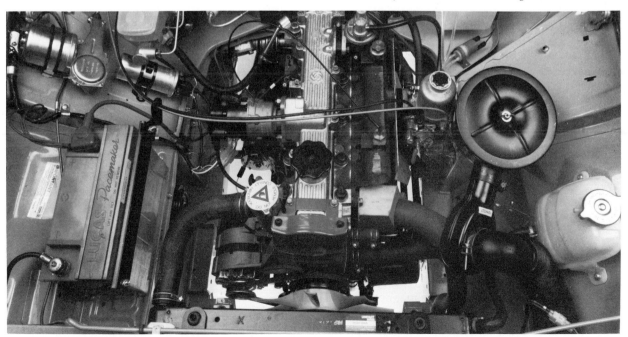

Questions on chapter 11

1. i Describe, with the aid of a labelled diagram, the structure and action of a simple two-pole electric motor. Explain the action of the commutator.
ii Will the motor run on direct current only, on alternating current only or on both? Give reasons.
iii Explain how the direction of rotation of the motor may be reversed. (E.A.E.B.)

2.

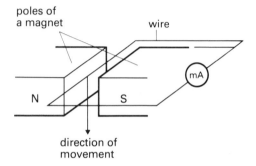

poles of a magnet · wire · mA · N · S · direction of movement

If the wire is moved in the direction shown, the milliameter will indicate a current flow.
  i State two factors which govern the size of the current.
  ii State two ways by which the direction of the current flow could be reversed. (W.M.E.B.)

3.

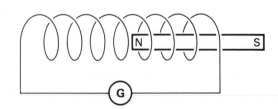

The diagram above shows a bar magnet being used to induce a current in a coil. The magnet is moving slowly into the coil and the galvanometer **G** is recording a steady current.
Describe any changes that occur when
  i The *speed* of the magnet is increased:
  ii A coil with *more turns* is used instead;
  iii The *direction* of motion of the magnet is *reversed*;
  iv The magnet *stops* moving.

4.

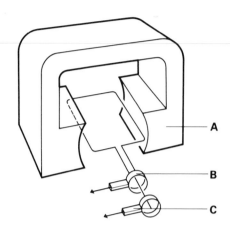

The diagram shows a simple A.C. generator.
  i Name the parts marked **A**, **B**, and **C**.
  ii Sketch a graph of the output voltage (Y-axis) against time (X-axis) for the generator and mark on it one position where the coil is vertical, and one where it is horizontal.
  iii Sketch the output voltage/time graph if a split ring commutator is used.
  iv Suggest one reason why electro-magnets are preferred to permanent magnets for most generators. (W.M.E.B.)

5. The diagram shows a moving coil meter.

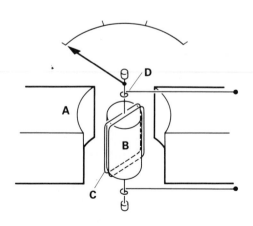

  i Name the parts labelled **A**, **B**, **C** and **D**.

ii

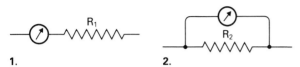

1.　　　　　　**2.**

Copy out and complete the following sentences:
In diagram **1** the resistance R_1 is said to be
in * and makes the meter into a *. It is
* compared with the resistance of the meter.
In diagram **2** the resistance R_2 is said to be in *
and makes the meter into a*. It is * compared
with the resistance of the meter.　(A.L.S.E.B.)

6. A milliammeter gives full-scale deflection when
a current of 10 mA passes through it, and it has a
resistance of 10 ohms.
　i What will be the potential difference across it
at full-scale deflection? It is desired to convert
this meter into an ammeter reading up to 1A.
　ii Should an additional resistance be placed in
series with the meter, or in parallel with it?
　iii What will be the current through this
additional resistance when the meter is showing
full-scale deflection?　(W.M.E.B.)

7.

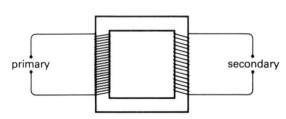

The diagram shows a simple transformer.
　a Copy out the following sentences and
complete them.
　　i The input to the primary coil must be an *
current.
　　ii The output current from the secondary coil
can be * than the input to the primary.
　　iii In the diagram the transformer is a step *.
　b If the mains voltage of 250 volts is put across
the primary coil of a transformer with 1 000
turns on the primary coil and 100 turns on the
secondary coil, what will be the output
voltage?　(A.L.S.E.B.)

8. **i** Why is electricity transmitted at such high
voltages on the National Grid system?
　ii Why is it not used at such high voltages in
our homes?
　iii Draw a labelled diagram of a step-down
transformer.　(W.M.E.B.)

9.

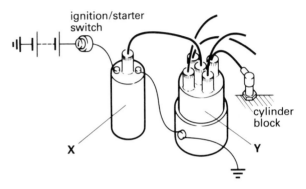

The diagram shows part of the high tension circuit
of a motor car. Name the parts marked X and Y and
explain their function.　(W.M.E.B.)

10.

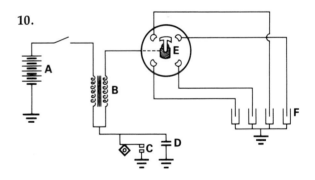

　a The diagram shows an electric circuit
commonly used in a car.
　　i Which electrical system is it?
　　ii Name the parts labelled **A** to **F**.
　b **i** Are the lights of a car wired in series or
parallel?
Give a reason for your answer.
　　ii A car has two headlights each rated at
12 V 48 W, two tail and two side lights each
rated at 12 V 6 W and a number plate light
rated at 12 V 6 W. Calculate the total current
taken by these seven lights from the car
battery.　(N.W.R.E.B.)

Electron beams

Electrons can be released from a metal surface, and made to travel in narrow beams. Some properties of these electron beams need to be explained to understand the ways they are used in many devices, including television.

Thermionic emission of electrons
Any hot piece of metal will give off, or *emit,* electrons. The electrons stay in a "cloud" around the metal. In practice an electrically heated thin coil of metal wire, called a *filament,* is used. The filament is usually made of tungsten as this readily emits electrons. A high filament temperature causes a large emission of electrons:

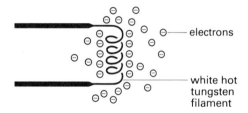

The diode
This device will only allow electricity to pass through it in one direction. It consists of a heated filament which gives off electrons, and a metal plate. The two are mounted in a glass bulb from which all the air has been removed:

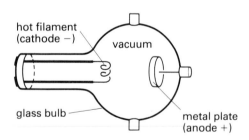

If the metal plate is made positive and the filament negative, the electrons will be attracted to the metal plate and a current flows, as in diagram **1**:

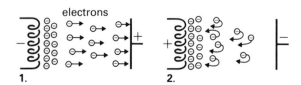

1.   2.

But if the metal plate is made negative and the filament positive, no current flows. The electrons are repelled from the metal plate and attracted back to the filament, as in diagram **2**. The metal plate is called an *anode* because it has to be positive before a current will flow. The filament is negative and is called a *cathode.*

The electron gun
This device, shown in the diagram, is used to make a beam of electrons:

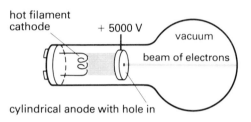

The electrons emitted by the cathode are attracted to the anode. By the time the electrons reach the anode, they are going so fast that some of them pass straight through the hole without being stopped. In this way a beam of electrons is produced. *Cathode rays* is the old fashioned name given to a beam of electrons.

Properties of electron beams
In order to understand how electron beams are used it is necessary to understand their properties – and also how they are controlled.

Fluorescence. When a beam of electrons strike certain substances light is given off. The colour of the light depends on the material. This effect is called *fluorescence,* and the materials are *fluorescent.*

Straight line motion. Electrons travel in straight lines. This can be demonstrated using the apparatus shown below, and in the photograph:

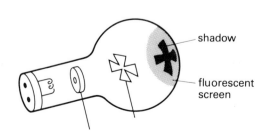

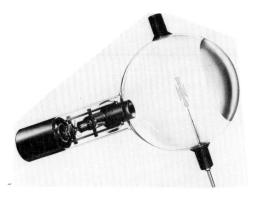

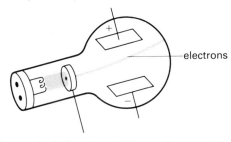

The fluorescent screen glows where the electrons strike it. There is a sharp shadow of the cross where the electrons are prevented from striking the screen. This suggests that electrons travel in straight lines.

Electric deflection. A beam of electrons is bent (deflected) when it is passed between two metal plates that have a potential difference between them. The beam is deflected towards the positive plate – electrons are attracted by the positive plate and repelled by the negative plate:

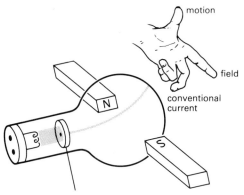

Magnetic deflection. When a beam of electrons is passed through a magnetic field it is also deflected. The direction is given by Fleming's left hand rule, as shown below. It must be remembered that the conventional current flows in the *opposite* direction to the electrons:

Controlling the strength of an electron beam

It is often necessary to control the strength of the electron beam produced by an electron gun. This is done by placing a mesh of thin wires called a *grid* between the anode and cathode of the electron gun. When the grid is at the same potential as the cathode – 0 V – it has no effect on the electrons, as shown in diagram **1** below.

When the grid is at −5 V as in diagram **2**, some of the electrons are reflected back to the cathode. The electron beam is weaker. If the grid is made still more negative, as in diagram **3**, all the electrons are reflected back to the cathode:

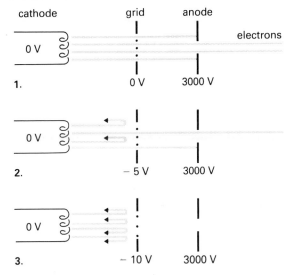

Exercises

Copy out the sentences in exercises 1 to 3 filling in the missing words from the lists.

1. Thermionic * is the * of * from a * metal.
(hot, boiling off, electrons, emission)

2. A diode is a * which conducts * in one * only.
(device, direction, electricity)

3. Fluorescence is the * of * from certain * when struck by *.
(materials, light, electrons, emission)

4. Describe two ways in which electron beams may be deflected.

5. Explain how the strength of an electron beam may be varied.

The cathode ray oscilloscope

The cathode ray oscilloscope can be used to measure potential differences or to show, in the form of a graph on a screen, how a potential difference varies with time.

The main part of the oscilloscope is the cathode ray tube:

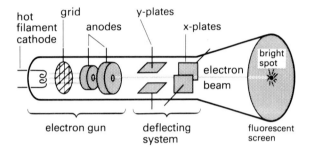

The tube has an electron gun which makes a beam of electrons. The screen at the end of the tube is coated with a fluorescent material so that a bright spot of light is produced where the end of the electron beam strikes it.

The brightness of the spot. This depends on the rate at which the electrons strike the screen. Making the grid less negative makes the spot go brighter.

Focussing the beam. Making a small, clear spot on the screen is done by altering the potential difference between the two anodes. The two anodes act as an "electric lens".

Deflecting the electron beam
Two sets of deflecting plates are used. The y-plates deflect the spot vertically. The x-plates deflect it horizontally.

y-plate deflection. A potential difference between the y-plates bends the electron beam. The spot on the screen moves up or down:

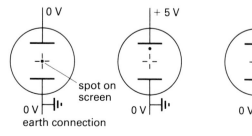

The cathode ray oscilloscope can be used as a voltmeter because the distance that the spot is deflected depends on the potential difference between the plates.

If an alternating potential difference from the mains is applied to the y-plates, the spot moves up and down 50 times a second making a straight line trace:

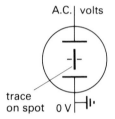

A straight line is seen because the screen continues to glow for a short length of time after the electron beam has moved on. The length of the trace is a measure of the maximum potential difference.

x-plates: time base. If the potential difference across the x-plates is increased steadily, the spot moves across the screen with constant speed. This is done by applying a potential difference which smoothly increases and then suddenly decreases:

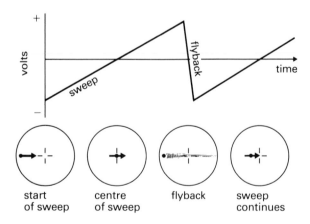

This is called a *saw-tooth* potential difference. On the sweep the spot moves at a constant speed across the screen to the right. On the "fly-back" the spot moves very quickly back to the left of the screen.

When a signal is applied to the y-plates, the time-base spreads it out along the x axis so that the variation of the signal with time can be seen. The next diagram shows how the time-base does this:

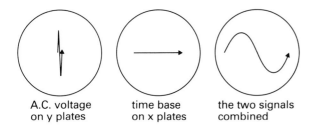

75

A.C. voltage on y plates | time base on x plates | the two signals combined

The controls of the oscilloscope
The photograph shows a typical cathode ray oscilloscope with the controls labelled.

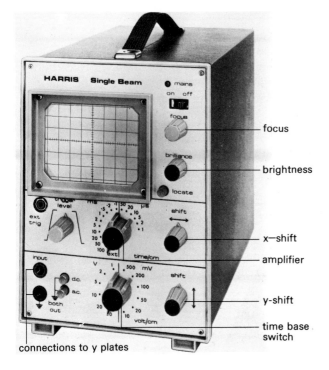

connections to y plates

Focus. This alters the potential difference between the two anodes.

Brightness. This makes the grid more or less negative.

x- and y- shift. These vary the 'no volts' across the deflecting plates so that the spot can be zeroed in the centre of the screen.

Time-base switch. This alters the length of the time the spot takes to move across the screen.

Amplifier. This magnifies the signal to the y-plates so that a small potential difference makes the spot move a large distance.

The black and white T.V. tube
This is a cathode ray tube with two time bases. One sweeps the spot across the screen, whilst the other pulls the spot down the screen. In this way the whole screen is illuminated with nearly horizontal lines. The T.V. signal from the aerial is applied to the grid of the tube to control the brightness of the spot. The different light and dark parts combine to form the picture.

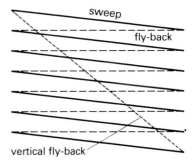

In Britain, T.V. screens have 625 horizontal lines. Colour televisions have a screen containing white, plus red, green and blue fluorescent pigments. One main electron gun activates the white pigment, three others activate the primary colours, which are superimposed on top of the black and white image. All the beams go to build up a complete coloured image.

Exercises
Copy out the sentences filling in the missing words from the lists.
1. In the cathode ray tube * are emitted from the *, accelerated by the * and strike the * screen making a bright spot of *.
(anode, cathode, light, fluorescent, electrons)

2. The brightness of the spot is controlled by making the * more or less negative, and the beam is focussed by * the * between the two *.
(altering, anodes, grid, potential difference)

3. Draw a labelled diagram of a cathode ray oscilloscope tube.
4. Explain how the cathode ray oscilloscope may be used as a voltmeter.

Solid state diodes and transistors

Substances are insulators or conductors depending on whether the electrons round each atom are bound to the atomic nucleus, or whether they are free to move. The element germanium is normally an insulator. It can be made to conduct by adding small quantities of impurities. This effect is used to make diodes and transistors.

'N' material and 'P' material

Adding a small quantity of arsenic to germanium frees some of the electrons so that the germanium will conduct in the normal way. This is called N-material because Negative electrons carry the current.

Adding a small quantity of indium to the germanium also makes it conduct – but this time, by the movement of "positive holes". A 'hole' is a place where an electron could be but is not. If an electron jumps into a hole it leaves behind another hole which can then be filled by another electron. In this way the holes move in the opposite direction to the electrons and behave like Positive charges. This is called P-material.

The junction diode

This consists of a piece of N-material and a piece of P-material joined together. If a battery is connected as shown below, the free electrons are attracted to the right and the holes to the left. In the middle there is a layer which has no holes or free electrons – it is an insulator so no current flows:

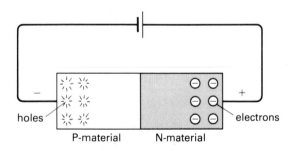

If the battery is connected the other way the electrons and holes flow across the junction – so a current flows as shown at the top of the next column.

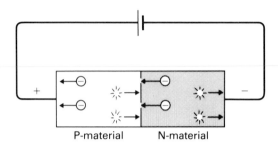

In this way, the junction diode conducts a current in only one direction. Its main use is for *rectification* – converting an alternating current into a direct current.

Rectification

Rectification can be demonstrated with a cathode ray oscilloscope. The circuit used to show the waveform of an alternating current is given below:

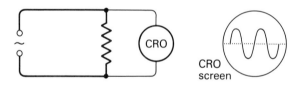

When a diode is placed in the circuit the current will flow in one direction only. The diode cuts off the negative half of the alternating current cycle:

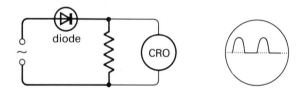

One use of the diode is in a car battery charger. The circuit consists mainly of a step-down transformer and a diode:

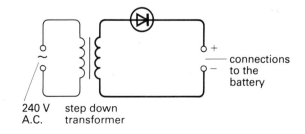

76

The step down transformer reduces the potential difference of the mains from 240 V to about 15 V. The diode changes the alternating current to a direct current which can be passed in one direction through the battery to charge it.

The transistor

There are two types of transistor; the N-P-N transistor and the P-N-P transistor. They consist of three layers of germanium:

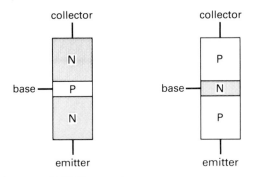

The first layer is called the *emitter,* the middle layer the base, and the third layer the collector. The two types of transistor behave similarly. The N-P-N transistor is described here as it is now the more common type.

The transistor is important because a small change in a current flowing from the base to the emitter causes a large change in current flowing from the collector to the emitter:

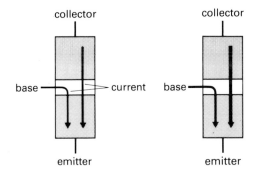

The circuit diagram shows how the transistor can be used as an amplifier to make the small varying current from a microphone large enough to work a loudspeaker:

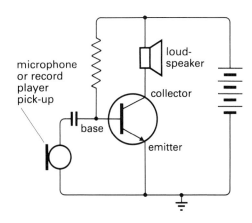

The microphone makes a small varying current which flows from the base to the emitter. This makes a large varying current flow through the transistor from the collector to the emitter and through the loudspeaker.

10-transistor circuit in a modern radio

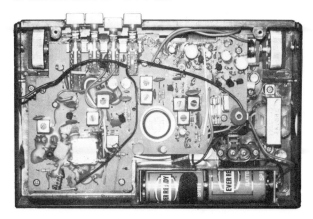

Exercises

1. What is meant by N-material and P-material?
2. Explain how a junction diode works.
3. What is meant by rectification? Draw diagrams to show an AC waveform and the rectified waveform.
4. Draw the circuit diagram for a simple battery charger.
5. Draw a schematic diagram of an N-P-N transistor. Label the N-material, P-material, base, emitter, and collector.
6. Draw a circuit diagram to show how an N-P-N transistor may be used as an amplifier.

Electrons and radiation

This section is about the photoelectric effect and X-rays. The photoelectric effect describes the way in which electrons are given off from certain substances when infra-red visible light, or ultra-violet radiation strikes them. X-rays are produced by the opposite effect, when fast moving electrons strike a substance.

The photoelectric effect

The photoelectric cell shown in the diagram makes use of this effect to produce an electric current which depends on the intensity of light falling on the cathode:

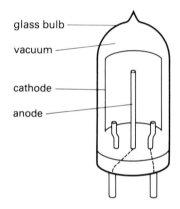

This type of cell can be used to make a light meter when it is connected in a circuit as shown below:

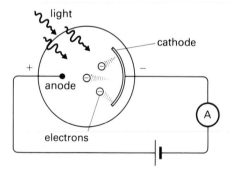

When the light strikes the cathode electrons are ejected from the surface by the photoelectric effect. The electrons are repelled from the negative cathode and attracted to the positive anode so that a current flows round the circuit.

Burglar alarm. The photoelectric cell is operated by infra-red radiation, which is invisible. When the

infra-red ray is broken, electrons are no longer emitted from the cathode of the photoelectric cell:

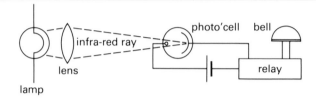

The current stops flowing through the relay, and this switches on the bell. The relay is designed so that the bell continues to ring until it is switched off.

Sound track on a film. The transparent sound track down one side of a film lets varying amounts of light through to the photoelectric cell. This makes a varying electric current which is amplified and fed to a loudspeaker to produce the sound:

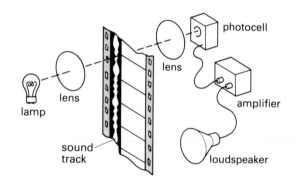

X-rays

X-rays are produced whenever fast moving electrons strike a metal and are stopped or slowed down:

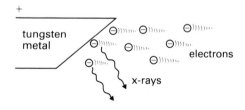

The Coolidge type X-ray tube. This type of X-ray apparatus is now generally used to produce X-rays:

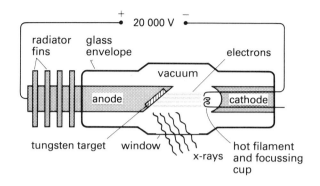

77

Electrons are emitted from the hot filament cathode, and then accelerated by the potential difference of 20 000 V to the target anode. As the electrons strike the target only 1% of their kinetic energy is converted into X-rays, and 99% is converted into heat. The radiator fins are to remove this heat. The intensity of the X-rays can be increased by making the filament hotter, so releasing more electrons to strike the target.

X-rays in medicine. X-rays penetrate matter. Materials of low density will not stop X-rays, but materials of high density will. Since they will also darken a photographic plate, they can be used to "see" inside the body. The X-ray photograph below shows a fractured arm joint. X-ray photographs are also used to detect chest and other diseases.

X-rays will pass through flesh – very high intensity X-rays also burn it. For this reason hospitals use them to treat cancer in the hope of burning the cancer cells inside the patient's body.

X-rays in industry. X-rays can be used to detect flaws in metal parts. Powerful two-million volt X-ray sets are used to examine the quality of metal castings, and welded joints. Gaps that are invisible on the surface can be easily detected.

Exercises

Copy out the sentences below filling in the missing words from the lists.

1. The photoelectric effect is the * of * from a * surface when *-red, visible or *-violet light strikes it.

(ultra, infra, metal, emission, electrons)

2. X-rays are produced when high velocity * strike a * and are * or * down.

(stopped, metal, electrons, slowed)

3. Explain how a photoelectric cell may be used to make a burglar alarm.

4. What does the sound track down the side of a movie film look like? Explain how it is used to produce the sound.

5. Draw a diagram of a Coolidge type X-ray tube and explain how it works.

6. Describe two uses of X-rays.

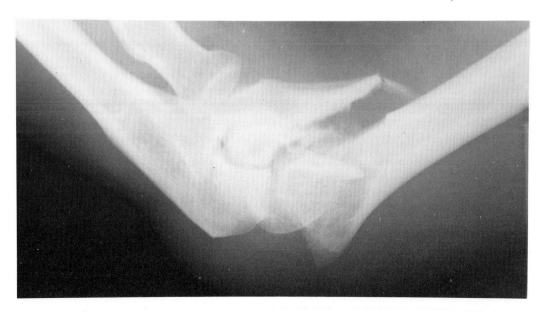

Atoms, isotopes and ionization

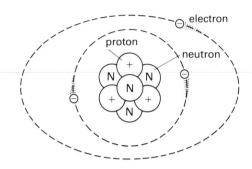

There are ninety-two naturally occuring simple substances which cannot be split up into simpler ones. These basic substances are called *elements* and all other substances are made by combining these elements in different proportions in a chemical reaction.

A sample of a few grams of an element contains billons of tiny, exactly identical particles called *atoms*. An atom is the smallest particle that can take part in a chemical reaction. Each one is made up of three different particles:

The proton. This has a mass of about one billion billionth of a gram. Because it is so small this mass is simply called an *atomic mass unit* (AMU). Each proton also has a tiny, positive charge, called *one unit of charge*.

The neutron. This has a mass of 1 AMU, the same as the proton, but no charge.

The electron. This has a much smaller mass than either the neutron or the proton – it has a mass of about 1/2000 AMU. The electron has one unit of negative charge, which is equal and opposite to the charge on the proton. Like the AMU, the size of the unit of charge is very small – about 7 million billion electrons have to flow past one point in a wire every second to make a current of 1 amp.

A model of the atom

The mass of the atom is concentrated in a very small volume at its centre which is called the *nucleus*. The nucleus has a diameter of about a billionth of a millimetre and is made up of neutrons and protons. The electrons are a long way from the nucleus and move around it in circular paths, rather like the planets around the sun. The diameter of the outer electron orbit – the diameter of the atom – is a million times bigger than the diameter of the nucleus.

In an atom there are equal numbers of electrons and protons. The negative charge on the electrons cancels out the positive charge on the protons, which makes the atom electrically neutral.

The diagram at the top of the next column shows an atom of the element *lithium*. This example is used to introduce the words used to describe atoms. You can see that the lithium atom contains three protons, three electrons and four neutrons:

Atomic number Z. This is the number of protons in the nucleus. Lithium has an atomic number of 3.

Neutron number N. This is the number of neutrons in the nucleus. The neutron number of lithium is 4.

Atomic mass number A. This is the number of protons plus neutrons in the nucleus. Lithium has an atomic mass number of 7.

$$\begin{matrix} \text{atomic mass} \\ \text{number} \end{matrix} = \begin{matrix} \text{atomic} \\ \text{number} \end{matrix} + \begin{matrix} \text{neutron} \\ \text{number} \end{matrix}$$

$$A = Z + N$$

All the details of the lithium atom can be represented symbolically as shown below:

$^{7}_{3}\text{Li}$ $\begin{smallmatrix}\text{atomic mass}\\\text{atomic number}\end{smallmatrix}$ Chemical symbol

Isotopes

Isotopes of an element are atoms having the same atomic number but a different mass number. The diagram below shows the three isotopes of hydrogen:

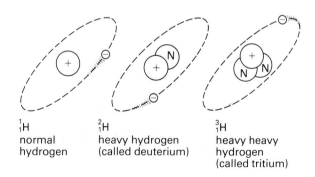

$^{1}_{1}\text{H}$
normal hydrogen

$^{2}_{1}\text{H}$
heavy hydrogen (called deuterium)

$^{3}_{1}\text{H}$
heavy heavy hydrogen (called tritium)

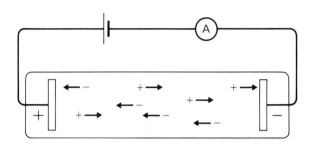

78

Each nucleus has one proton (same atomic number) but different numbers of neutrons and so the mass number is different. Only the isotopes of hydrogen are given separate names. In general isotopes are indentified by quoting the mass number after the name of the element, e.g. uranium 238 ($^{235}_{92}$U) or or uranium 235 ($^{238}_{92}$U)

Nuclide. The term isotope is gradually being replaced by the more modern term nuclide. A nuclide is any atom which has a characteristic number of neutrons and protons.

Atomic mass unit. This is specially chosen so that the mass of the nuclide carbon 12 ($^{12}_{6}$C) is 12 atomic mass units. Carbon 12 is chosen as the standard rather than hydrogen 1 because it is easier to get carbon 12 without any other nuclides mixed with it.

Ionization in gases

An atom is normally electrically neutral. The negative charge on the electrons is cancelled out by the positive charge of the protons in the nucleus. An atom with one electron missing has a net positive charge. It is called a positive ion. The electron that has been removed from the atom is called a negative ion. The two are said to make up an ion pair. The diagram shows how an ion pair is made from a neutral helium atom:

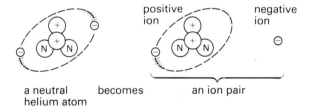

a neutral helium atom   becomes   an ion pair

Causing ionization. A flame causes ionization. The high temperature of the flame greatly increases the speed of the air molecules. The air molecules move so fast that when they collide they knock electrons off one another making positive and negative ions.

Radioactivity also causes ionization. When a radioactive particle strikes an air molecule, it knocks an electron off it making a positive and a negative ion.

Ions and electric fields. The diagram shows two metal plates with ionized gas between them:

The positive ions are attracted to the negative plate and the negative ions are attracted to the positive plate. The movement of ions between the plates allows an electric current to flow round the circuit. The size of the current depends on the number of ions between the plates. This effect can be used to detect radioactivity because radioactivity causes ionization.

Exercises
1. What are neutrons, protons, and electrons?
2. Draw a diagram of a berylium atom which has a nucleus containing 4 protons and 5 neutrons. Write down the symbol for this atom.
3. Copy out the sentences filling in the missing words from the lists.

    Isotopes of an * are * having the same * number but a different * number.
(atoms, atomic, element, mass)

    A nuclide is any * which has a * number of * and *.
(atom, neutrons, characteristic, protons)

4. What is an ion pair?
5. How are ions affected by an electric field?

Radioactivity — methods of detection

Radioactive particles cause ionization in gases, cause flashes of light to be given off from certain substances, and affect photographic film. All these effects are used to detect radioactivity.

The Geiger-Muller tube

The Geiger-Muller (G.M.) tube shown in the diagram detects radioactive particles by the ionization that they cause:

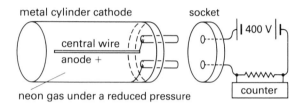

When a radioactive particle enters the G.M. tube through the mica "window" it makes a trail of ion pairs. The negative ions are attracted to the positive anode and the positive ions to the negative cathode. The movement of the ions allows a small pulse of current to flow round the circuit. These pulses are counted electronically by an instrument called a *scaler*:

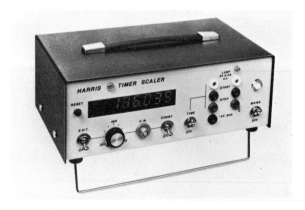

Background count

When the Geiger-Muller tube is connected to the scaler, one or two radioactive particles are recorded per second even though there is no radioactive source present. This is called the *background count*. The background count is produced partly by minute traces of radioactive substances in the air and materials of the building and partly by *cosmic rays*. Cosmic rays are very high energy radioactive particles which come from the depths of space.

The cloud chamber

This is a device which is used to make the tracks of radioactive particles visible. It does this by making droplets of liquid form along the trail of ions left by the radioactive particle. The liquid is usually alcohol.

At a particular temperature the air will only hold a certain amount of alcohol vapour. The air is then said to be *saturated*. The amount of vapour needed to saturate the air goes down as the temperature goes down. If the temperature is suddenly lowered, the air will contain more vapour than it should do and is said to be *supersaturated*. The excess vapour condenses out in the form of droplets— these droplets form most readily on gas ions.

The diagram below shows an expansion type cloud chamber:

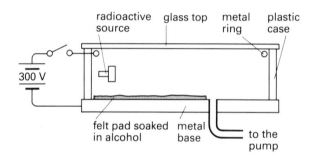

The cooling needed to make the air supersaturated is produced by reducing the pressure. The excess vapour condenses on the trails of ions left by radioactive particles making them visible as thin tracks of cloud. The next photograph shows some typical cloud chamber trails:

Before the chamber can be used again it has to be cleared of ions. This is done by applying a 300 V potential difference between the base and the metal ring at the top. The negative ions are attracted to the ring and the positive ions to the base.

Scintillation counter

Certain substances such as zinc sulphide emit flashes of light when radioactive particles strike them. These flashes of light are called *scintillations*. This effect can be observed in a dark room using the instrument shown below:

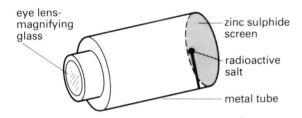

Early experimenters used to sit in dark rooms counting flashes of light. Today a scintillator is used in conjunction with a photoelectric cell. Each flash of light produces a pluse of electric current which is counted by a scaler.

Photography

All types of radioactivity blacken photographic films. The photograph below was produced by laying an electric drill on top of a film and placing a strong radioactive source one metre above it. The radiation penetrated the plastic case of the drill but not the solid metal parts.

Exercises

1. Draw a labelled diagram of the G.M. tube.

Copy out the sentences in exercises 2, 3 and 4 filling in the missing words from the lists.

2. When a * particle enters the G.M. tube it leaves a trail of *. The electrons are attracted to the * making a pulse of *, which is counted by a *.
(ions, current, radioactive, scaler, anode)

3. In a * counter a radioactive * strikes a zinc * screen, which * a * of light.
(flash, sulphide, scintillation, particle, emits)

4. A radioactive * makes a * of ions in a * chamber. * vapour condenses on these * as *.
(droplets, particle, ions, alcohol, cloud, trail)

5. Draw a labelled diagram of the cloud chamber.
6. Why do people who work with radioactivity or X-rays, wear a badge containing a piece of unexposed film in a sealed packet?

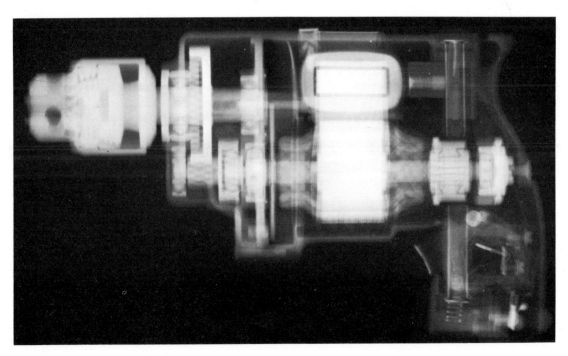

Radioactivity — properties

There are three different types of radiation emitted by radioactive materials.

Alpha (α) particles. An α-particle is the same as the nucleus of a helium atom. It contains two protons and two neutrons, has two units of positive charge and a mass of 4 atomic mass units.

Beta (β) particles. A β-particle is an electron. It has one unit of negative charge and a mass of about 1/2000 atomic mass units.

Gamma (γ) rays. γ-rays are electromagnetic radiation of a very short wavelength. They carry no electric charge and travel at the speed of light.

Properties of radioactive particles

Five different properties of radioactive particles are compared in the next paragraphs.

1. Range in air. α-particles and β-particles are slowed down and stopped by collisions with the air molecules. γ-rays are hardly affected by the air. Only their intensity gets less as they spread out:

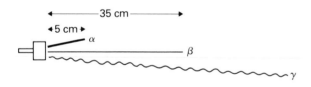

2. Absorption. The diagram shows the thickness and type of material which is needed to absorb each type of radiation:

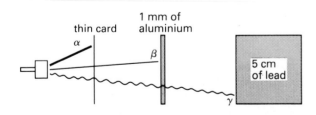

3. Effect of an electric field. The positive α-particles are attracted towards the negative plate, and the negative β-particles are attracted towards the positive plate. As the γ-rays have no charge they are not affected by electric fields:

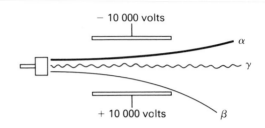

4. Effect of a magnetic field. The α-particles and β-particles are deflected in a direction which is at right angles to both the magnetic field and their direction of travel. They are deflected according to Fleming's left hand rule. In the diagram the α-particles, which travel in the same direction as the conventional current, are deflected upwards. The β-particles are deflected in the opposite direction, straight downwards. γ-rays are not affected by magnetic fields and carry straight on:

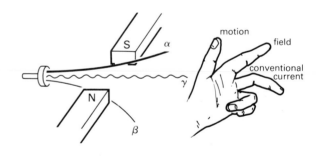

5. Cloud chamber trails. The diagram below shows cloud chamber trails produced by the three different types of radiation:

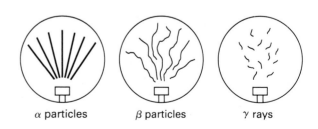

α particles   β particles   γ rays

α-particles make thick trails because they are good ionizers. The trails are straight because the α-particles are heavy and not easily deflected by collisions with air molecules.

β-particles make thin trails because they are poor ionizers. The trails are wavy because β-particles are light and easily deflected by collisions with air molecules.

γ-rays are very poor ionizers. The faint wavy trails are caused by electrons which have been knocked off by air molecules, when they have been struck by γ-rays.

Radioactive decay

When a nucleus emits a radioactive particle, it changes into another nuclide. This process is called radioactive decay. The rate of radioactive decay, which is measured by the number of radioactive particles emitted per second, is called the *activity*. The activity depends on the number of undecayed atoms present. The activity gets less as the number of undecayed atoms gets less.

The diagram shows what happens as a sample of 8 g of the radioactive gas radon 220 decays by emitting α-particles:

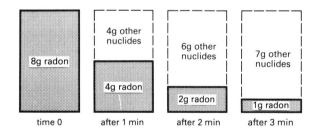

| 8g radon | 4g other nuclides / 4g radon | 6g other nuclides / 2g radon | 7g other nuclides / 1g radon |
| time 0 | after 1 min | after 2 min | after 3 min |

After 1 minute only 4 g of radon remains. The rest of the radon has decayed to form other nuclides. After the second minute, 2 g of radon remains – after the third minute, only 1 g of radon remains. Every minute, the amount of radon left and also the activity of the radon decreases by half.

Half life

The time taken for the activity of a substance to decay to half its original value is called the *half life*. The last diagram suggests that the half life of radon is one minute.

Measurement of half life. This is done by finding how the activity of a radioactive substance changes as time passes. A Geiger counter may be used to measure the activity and a graph of activity versus time is plotted. The graph below shows the sort of result that might be obtained:

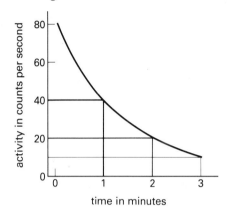

The half life is the time taken for the activity to decay to half its original value ie. from 80 to 40. The half life is seen to be 1 minute. It takes a further one minute for the count to drop from 40 to 20.

Carbon dating

Living plants continually absorb carbon dioxide from the air that contains a small proportion of radioactive carbon 14. One gram of carbon in a living plant has an activity of 15 counts per second. When the plant dies no more radioactive carbon is taken in. The radioactive carbon present decays very slowly, with a half life of 5600 years. Therefore a piece of wood which was found to have an activity of 7.5 counts per second would be about 5600 years old.

Exercises

1. Describe five properties each of α-particles β-particles and γ-rays.
2. What is meant by radioactive decay, activity, and half life?
3. A radioactive element has a half life of 2 days. If you start with 20 g, how much would you have left after 6 days?

Nuclear energy

When a neutron strikes a uranium 235 nucleus it releases two smaller parts, plus *three* more neutrons, and a lot of energy. This process is called *nuclear fission*:

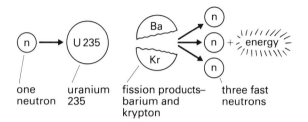

one neutron | uranium 235 | fission products– barium and krypton | three fast neutrons

The total mass of the fission products is *less* than the mass of the original uranium 235 nucleus. The reduction in mass is accounted for by the energy given off. The energy is released according to Einstein's famous equation:

$$\text{energy} = \frac{\text{loss of mass}}{\text{mass}} \times \text{velocity of light squared}$$

$$\mathbf{E} = mc^2$$

Each time a uranium nucleus splits, it releases three neutrons. These three neutrons may cause three further nuclei to split – the total of nine neutrons from them may cause nine more atoms to split up – an increasing *chain reaction* builds up:

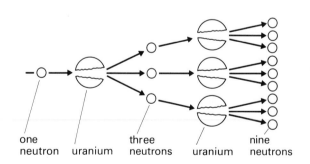

one neutron | uranium | three neutrons | uranium | nine neutrons

The nuclear (atomic) bomb

The result of an increasing chain reaction in uranium 235 is shown in the photograph. The reaction will only increase to make a huge explosion if the piece of uranium is made sufficiently big that most of the neutrons produced in a fission hit other uranium atoms before they escape.

The reaction will not increase in a small piece of uranium, because most of the neutrons can escape easily. In between these two amounts there is a *critical mass* where exactly one neutron from each fission comes to hit another atom. In this case the reaction will continue at a constant rate.

A simple nuclear bomb can be made as shown in the diagram. Two hemispheres of uranium 235, each of ¾ of the critical mass are separated in a tube, as in diagram **1** below. When the chemical charge is exploded the two hemispheres are forced together making one sphere of 1½ times the critical mass, as in diagram **2**.

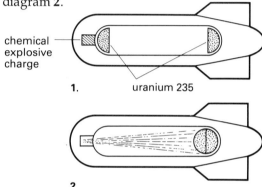

chemical explosive charge

1.   uranium 235

2.

The result is an increasing chain reaction and a nuclear explosion, as shown in the first paragraph.

The original problem in making the nuclear bomb was in obtaining the uranium 235. Natural uranium consists of a mixture of 99% uranium 238 and 1% uranium 235. These two isotopes are very difficult and costly to separate.

The nuclear reactor

A steady chain reaction which takes place at a constant rate as shown below, is the basis of the nuclear reactor:

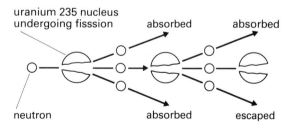

uranium 235 nucleus undergoing fisssion | absorbed | absorbed

neutron | absorbed | escaped

The reactor consists of a core, which contains natural uranium, *fuel rods* (99% uranium 238 and 1% reactive uranium 235), and cadmium *control rods* that absorb neutrons. Both are mounted in graphite, which acts as a *moderator* – it slows the neutrons down without absorbing them. The uranium 235 gives off neutrons, and the cadmium absorbs them – by gradually removing the cadmium rods, the reactor is started. The number of cadmium rods is very carefully controlled so that exactly the right number of neutrons are free to react. A bank of cadmium rods is ready to be dropped into the reactor, to quickly shut it off in case of emergency. The diagram below shows the core of a nuclear reactor:

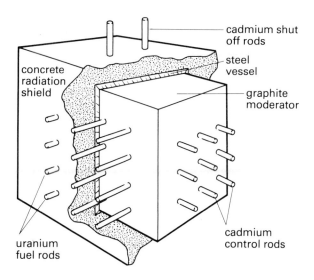

A British nuclear test. Most countries, including Britain, no longer test nuclear devices above ground.

The fission reaction produces heat energy. In a nuclear power station the reactor core replaces the coal or oil fired boiler in an ordinary power station. The diagram below shows how the heat from the reactor core is used to boil water. The steam produced drives a steam turbine which drives the electrical generator:

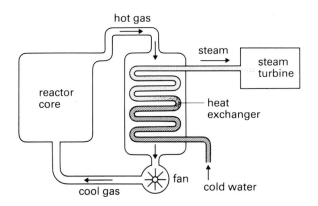

Exercises

Copy out the sentences in exercises 1, and 2 filling in the missing words from the lists.
1. Nuclear * is said to occur when a * strikes a * and splits it up into two * parts with the release of *.
(nucleus, large, fission, energy, neutron)

2. A * which slows down neutrons without * them is called a *.
(moderator, substance, capturing)

3. Explain how a nuclear bomb works.
4. Draw a labelled diagram of a reactor core. Explain the purpose of the moderator, and the cadmium control rods.
5. Explain how the heat from a nuclear reactor is used to produce electricity.

Questions on chapter 12

1.

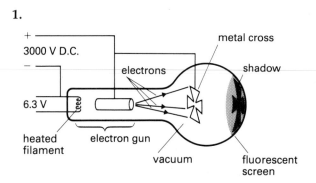

In the 'Maltese Cross' experiment above, an electron beam (cathode rays) cast a sharp shadow of a metal cross on the fluorescent screen.

  a What does this shadow tell us about the way in which cathode rays travel?

  b What is the function of the heated filament?

  c Why must a high voltage be applied across the electron gun?

  d i What would you expect to happen to the shadow on the screen if a positively charged rod were placed above the tube?

    ii Give a reason for your answer. (M.R.E.B.)

2.

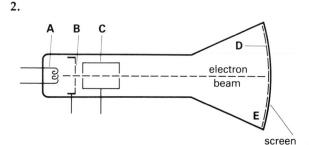

  i Copy the diagram of a simplified cross section of a cathode ray tube into your book and label the three components **A**, **B**, and **C**.

  ii State which is connected to the positive of a high tension supply.

  iii What is the purpose of the coating **D**?

  iv Also on your diagram, draw and label the parts which make it possible for electrostatic forces to deflect the beam of electrons towards **E**.

  v Draw a square to represent the screen of a television tube, and in it draw a diagram to show what is meant by the "Scanning action".

                    (Y.R.E.B.)

3.

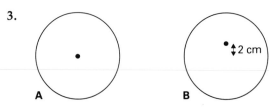

Diagram **A** shows the face of an oscilloscope tube with the time-base switched off and the voltage sensitivity set at 5 V/cm.

  i What is the input voltage shown in diagram **B**?

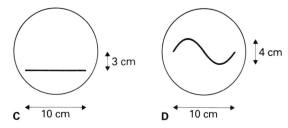

In diagrams **C** and **D** the voltage sensitivity is unaltered, but the time base has been set to one hundredth of a second per centimetre.

  ii Describe the input voltage shown in diagram **C** as fully as you can.

  iii Describe the input voltage shown in diagram **D** as fully as you can. (W.M.E.B.)

4. i Explain in simple terms the construction of a semi-conductor diode.

  ii Explain how this construction leads to the particular property of such a diode.

  iii Draw a circuit showing a semi-conductor diode being used to perform half-wave rectification.

  iv Sketch the output of your circuit. (W.M.E.B.)

5.

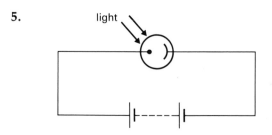

  i Explain briefly why current flows when light is shone onto the photo-electric cell.

  ii What difference does it make if a brighter light is used?

iii What difference does it make if light of a very much longer wave-length is used?
iv Suggest one use of a photo-cell. (W.M.E.B.)

6. i Draw a diagram of a simple modern X-ray tube.
ii State two effects of X-rays.
iii Why should care be exercised in the use of X-rays?
iv Give one way in which the danger may be reduced. (W.M.E.B.)

7.

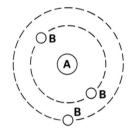

The diagram represents a model of an atom.
i Name the part labelled **A**.
ii Name two particles contained in **A**.
iii Name the particles labelled **B**.
iv What type of force exists between **A** and the particles labelled **B**. (W.J.E.C.)

8. a Copy and complete the following table.

particle	mass	charge
proton		
neutron		
electron		

b How are these particles arranged in the atom?
c Draw and label a simple picture of a helium atom that has an atomic number of 2 and a mass number of 4.
d Natural chlorine consists of two types of atoms both of atomic number 17, but one has a mass number of 37 and the other 35.
i What name do we give to atoms which differ in this way?
ii What accounts for the differences in the atoms?
iii In what ways are the atoms similar?
(A.L.S.E.B.)

9. a Explain what is meant by atomic number and mass number.

b $^{14}_{6}C$ is a radioactive form of carbon.
i How many protons are there in the nucleus?
ii How many neutrons are there in the nucleus?
c What is an ion? (E.A.E.B.)

10.

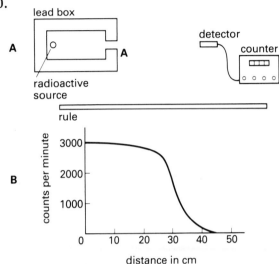

The apparatus shown is used to measure the range of travel of beta particles in air. The graph shows typical results. The source of beta particles is strontium 90 with a half life of 20 years.

i Why is the lead box used?
ii What kind of charge has a beta particle?
iii What is the approximate mass of a beta particle compared to the mass of the hydrogen atom?
iv Name a suitable detector which could be used.
v How would the counting rate of the counter be affected if a sheet of paper were held vertically at **A**?
vi How would the counting rate be affected by a strong magnetic North Pole at **A**?
vii If the present count rate is 2 000 counts per minute, how long will it take to fall to 500 counts per minute?
viii Use the graph to find how far the average beta particle travels through air. (Y.R.E.B.)

11. Diagrams **A** and **B** show tracks caused by two different radioactive sources placed in a cloud chamber, viewed from above.

a i Describe how a cloud chamber works and what these tracks are.

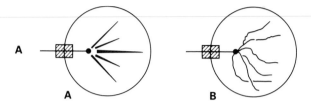

A

A   B

ii State the kinds of particles which are causing the tracks in **A** and **B**.

b What might you expect to observe if
i A piece of paper were placed in front of the source in each case?
ii A magnet were held alongside the chamber as shown in Diagram **C**?

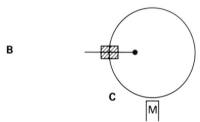

B

C

M

iii A gamma ray source were put in the cloud chamber instead of the other sources?
(M.R.E.B.)

12. A geiger tube samples some radioactive Radon gas, an alpha particle emitter, every 30 seconds. The results are given in the table below. Background count has been allowed for.

time in seconds	count rate
0	8040
30	5040
60	3360
90	2240
120	1440
150	880
180	600
210	400
240	240

a Draw a graph of count rate (vertical axis) against time.

b From your graph determine the time when:
i the count rate was 6 000,
ii the count rate was 3 000.
c i Calculate the time which elapsed between the 6 000 and 3 000 count rates. (Use your answers to **b** above).
ii What is the significance of this time?
d What is meant by the term *background count,* and how is it caused?   (E.A.E.B.)

13. a i What is a neutron?
ii Why is it particularly valuable for bombarding atomic nuclei?
iii Why is energy released when a uranium 235 nucleus breaks up into two smaller nuclei?
iv What is the name given to this break-up?
v What is meant by a chain-reaction, and why is a "critical mass" necessary to sustain one?

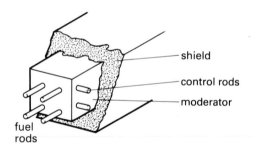

shield

control rods

moderator

fuel rods

b The diagram shows the basic arrangement of a nuclear reactor. Explain the functions of the control rods, the moderator and the shield. Why does uranium 238 not make a suitable fuel for reactors?   (W.M.E.B.)

Acknowledgements

The Publisher would like to thank the following for permission to reproduce photographs:

Associated Press, figs. 35.3, 39.1
Austin Morris, figs. 72.2, 73.2, 73.3
Mark Berry, fig. 9.1
B.I.C.C. Ltd., fig. 11.2
British Broadcasting Corporation, fig. 29.1
British Oxygen, fig. 18.3
British Rail, fig. 28.2
British Steel Corporation, fig. 65.1
Building Research Establishment, fig. 23.4
Camera Press, figs. 13.2, 16.1, 29.2, 38.1, 43.1, 45.3
Electricity Council, fig. 62.1
Dr. Elliott, Imperial College, fig. 79.2
Dr. Faulkner, Radcliffe Hospital, fig. 77.1
Fibreglass Ltd., figs. 31.1, 31.5
Peter Fraenkel, fig. 11.1
General Electric Company Ltd./Hotpoint, fig. 24.2
Gallenkamp, Griffin and George, figs. 4.2, 18.2, 24.1, 50.2, 50.3, 60.1, 64.2, 71.1, 72.3
Gilbey Vintners, fig. 12.1
Philip Harris Ltd., figs. 40.1, 67.1, 68.1, 71.2a, 71.b2, 71.3, 75.1, 79.1
Kelvin Hughes, fig. 54.1
Ralph Morse/LIFE © Time Inc. 1979 fig. 33.1
London Transport, fig. 2.1
Motor Magazine, figs. 4.1, 11.3, 12.1
Monsanto Chemicals, fig. 5.1
Natural History Photographic Agency, fig. 23.2a
NASA, figs. 1.1, 1.2, intro.
Oxford Mail, fig. 3.3
Perkins Engines, fig. 37.1
Picturepoint, figs. 1.3, 2.2, 49.1, 50.1, 70.2
Popperfoto, fig. 7.1
P.S.C. Equipment Ltd., figs. 20.1, 28.1b
Rolls-Royce Ltd., fig. 35.4
Salters Housewares Ltd., 3.1
Shell (UK) Ltd., fig. 23.2b
Smiths Industries, figs. 18.3a, 72.1
Space Frontiers Ltd./NASA, fig. 35.1
Stanley Gibbons, fig. 3.4
Stanley Tools, figs. 14.1, 14.2
Survey and General Instruments Ltd., fig. 18.1
Syndication International, figs. 6.1, 12.2, 15.1
Teltron Ltd., fig. 74.1
Transport and Road Research Laboratory, fig. 10.1

United Kingdom Atomic Energy Authority, figs. 35.5, 79.3, 81.1
Wimpey Ltd., fig. 28.1a
C. Roger Wood, fig. 15.2

Cover illustrations by kind permission of the United Kingdom Atomic Energy Authority and Ann Ronan Picture Library.

The publishers also wish to acknowledge the assistance of the C.S.E. boards who have granted permission to reproduce questions from their examinations. Individual questions are credited by initial to the board concerned.

Answers to numerical questions

Exercises

3	3	3 N
5	3	aluminium d = 2.5 g/cm³
		lead d = 11.0 g/cm³
6	3	RD of hexagonal object = 3.5
		RD of square object = 3
	4	RD of liquid = 0.8
	5	RD of object = 7 RD of liquid = 1.25
8	3	25 m/s
	4	10 m/s²
	6	20 m
9	5	**a** 3 m/s² **b** 1.5 m/s² **c** 1350 m
10	3	5 m/s²
	4	652 N
	5	4 kg
11	2	WD = 2 500 J Power = 500 W
	3	WD = 1 800 J Power = 450 W
13	4	VR = 3 MA = 2 Efficiency = 0.75 or 75%
14	2	Gear B rotates clockwise at a speed of 40 revs/min VR = 0.5
	3	VR = 5 MA = 4 efficiency = 0.8 or 80%
	4	VR = 4 MA = 3 efficiency = 0.75 or 75%
16	1	64 000 N/m²
	3	d_1 = 800 kg/m³
	4	d_1 = 600 kg/m³
	5	depth = 20 m
17	4	vacuum space is about 1.24 m long
	5	atmospheric pressure = 108 800 N/m²
20	2	maximum force = 2 000 N
25	3	3 360 J
	4	42 000 J
	5	192 000 J
	6	640 J/kg K
	8	840 J/kg K
26	3	33 600 J
	4	22 600 J
	5	34 000 J
	6	378 000 J/kg
	7	2 310 000 J/kg
27	1	1.5 m³
	2	100 N/m²
	3	75 N/m²
	4	600 K
	5	8 m³
	6	600 K
	7	33.3 N/m²

30	3	maximum 30 °C, minimum 2 °C, actual 12 °C
42	4	1.60
	6	1.5
47	1	**a** 1 cm high, real, 4.5 cm from lens
		b 2 cm high, real 6 cm from lens
		c 3 cm high, real 7.5 cm from lens
		d 4 cm high, vertical 3 cm from lens
	2	1.33 cm high, 2.67 cm from lens
	3	**a** 1 cm high, real 6 cm from mirror
		b 2 cm high, real 8 cm from mirror
		c 4 cm high, real 12 cm from mirror
		d 4 cm high, vertical 4 cm from mirror
	4	1.5 cm high and 1.88 cm from mirror
54	2	311 m/s
	3	2 km
	5	217 m
	6	644 Hz 3450 m/s 2 m
60	3	5 Ω
	4	0.5 A
	5	6 Ω 8 V across 4 Ω resistor
		4 V across 2 Ω resistor
	6	1 Ω
61	2	5 A; 10 A; 2 A
	3	2300 J/kg K.
62	5	**a** 24 p **b** 1.5 p **c** 2 p **d** 6 p
	6	**a** 3.24 p **b** 27 p **c** 10.8 p
80	3	2.5 g

End of Chapter Questions

Chapter 1:
Forces

2	**ii** 50 N
3	1 m
4	**iii** 8 cm **iv** 30 N
5	**a i** 200 cm³ **ii** 0.5 g/cm³ **iii** yes
	iv 100 cm³ **iv** half
	b i 1.5 g/cm³ **ii** no
6	**i** 4 N **ii** 4 N **iii** 400 g
	iv 400 cm³ **v** 2.5 g/cm³
7	**i** 2 048 kg **ii** 20 480 N
	iii 20 480 N **iv** 39 520 N
9	**ii** 1 225 kg/m³

Chapter 2:
Forces on the move
1   **b** 100 m   **c** 10 m/s   **d** 2 m/s²
3   **i** **a & b** 2 m/s   **ii** **c & d** 1 m/s
     iii **b & c** 250 s   **iv** **e & f**   **v** 100 m
     vi 100 m   **vii** 500 s
4   **c** 17 s & 43 s   **d i** 2 m/s²   **ii** 1.5 m/s²
7   **i** 5000 J   **iii** 5000 J
8   **v** 3600 J   **vi** 180 W
9   **b i** 1500 N
10   MA = 3.33   VR = 4   efficiency = 0.833
11   **i** 5   **ii** 2   **iii** 0.4

Chapter 3:
Pressure
1   **ii** 1 600 N/m²
2   **ii** 6 m²   **iii** 6 000 N/m²
4   **ii** 960 mm of mercury
8   **v** 76 cm
9   **iii** 250 000 N/m²
10   **i** 4 N/m²   **ii** 8 N

Chapter 4:
Kinetic Theory
4   42 °C
7   **i** 25 °C
8   **c i** 120 000 J   **iii** 1 000 W   **iii** 2 000 W
9   30 J
10   **ii** 16 cm³   **iii** 184 kN/m²
11   **i** 10 N   **ii** 10 J   **iii** 1 000 J   **v** 780 J
     vi 0.78

Chapter 5:
Expansion and Transmission of heat
4   **c ii** 26.7 °C   **iii** 9 cm

Chapter 6:
Properties of light
7   **iii** 1.5
9   **v** 48°   **viii** 1.9

Chapter 7:
Lenses and curved mirrors
6   **iii** 33.3 mm
10   **vi** 4.5 mm   **vii** 1.5 cm

Chapter 8:
Vibrations and Waves
1   **i** 4 Hz   **ii** 0.5 m
2   **a ii** 200 000 Hz   **d** 0.5 s
8   2 250 m
10   **i** 50 cm   **ii** 2 Hz   **iii** 100 cm/s

Chapter 9:
Electricity
4   **c i** 12 V   **ii** 24 V
5   **i** 6 Ω   **ii** 12 V   **iii** 24 W
6   **c** 0.2 A   **d i** 4 Ω   **ii** 9 Ω   **iii** 1 A
     iv 0.67 A
7   **c** 6.6 Ω
8   **b i** 8 A
10   54 p
11   **i** 4 A   **ii** 5 A   **iii** £1.75
12   **ii** 0.25 A   **iii** 960 Ω

Chapter 11:
Making and using electricity
6   **i** 0.1 V   **iii** 0.99 A
7   **b** 25 V
10   **b ii** 10.5 A

Chapter 12:
Electron and Nuclear Physics
3   **i** 10 V
10   **vii** 40 years   **viii** 32 cm
12   **b i** 20 s   **ii** 68 s   **c** 48 s

Maths, units, and symbols

Turning equations round

In order to find the unknown quantity in a mathematical equation, it is frequently necessary to change the equation round. In the equation below the quantity x is needed:

$$\frac{P}{Q} = \frac{S}{x}$$

The equation must be put into the form:

$$x = \text{"something"}$$

There are simple rules for doing this:

1. Cross multiply to get everything on one line:

$$\frac{P}{Q} \diagdown = \diagdown \frac{S}{x}$$

$$\therefore \quad Px = QS$$

2. Put the things next to the x under the other side:

$$Px = QS$$

$$\therefore \quad x = \frac{QS}{P}$$

Example 1. Find V_2 in the general gas law equation:

$$\frac{P_1 V_1}{T_1} = \frac{P_2 V_2}{T_2}$$

1. Cross multiply:

$$P_1 V_1 T_2 = T_1 P_2 V_2$$

2. Put the things next to V_2 under the other side:

$$\frac{P_1 V_1 T_2}{T_1 P_2} = V_2$$

$$\therefore \quad V_2 = \frac{P_1 V_1 T_2}{T_1 P_2}$$

Example 2. Find I in Ohm's law:

$$V = IR$$

As the equation is already on one line rule 1 is not needed.

2. Put the things next to I under the other side.

$$\frac{V}{R} = I$$

$$\therefore \quad I = \frac{V}{R}$$

Exercises

1. Find λ in $\quad V = f\lambda$
2. Find I in $\quad P = VI$
3. Find t in $\quad H = RI^2 t$
4. Find S in $\quad H = m\, s\, (\theta_2 - \theta_1)$
5. Find T_1 in $\quad \dfrac{P_1 V_1}{T_1} = \dfrac{P_2 V_2}{T_2}$

Units

The first table given below is a list of all the "basic" units used in the study of physics

name	symbol	name	symbol
metre	m	watt	W
kilogram	kg	degree celsius	°C
second	s	degree kelvin	K
hertz	Hz	ampère (amp)	A
newton	N	ohm	Ω
joule	J	volt	V

Units for physical quantities

Quantities which have "per" in them (such as density: kilograms *per* cubic metre) may be written in two ways – as kg/m^3, or as $kg\, m^{-3}$. In this book, units are all written as the example kg/m^3.

physical quantity	unit
mass	kg
length	m
time	s
frequency	Hz
area	m^2
volume	m^3
density	kg/m^3 $(kg\, m^{-3})$
velocity, speed	m/s $(m\, s^{-1})$
acceleration	m/s^2 $(m\, s^{-2})$
force, weight	N
pressure	N/m^2 $(N\, m^{-2})$
work, energy, heat	J
power	W
temperature	°C or K
specific heat capacity	$J/kg\, K$ $(J\, Kg^{-1} K^{-1})$
specific latent heat	J/Kg $(J\, Kg^{-1})$
electric current	A
electrical resistance	Ω
electromotive force	V
potential difference	V

Multiples of units

Special words are put in front of the basic unit to represent bigger or smaller quantities. This is explained in the table below, using the example of the volt:

multiple	name	symbol
one million	*mega*volt	MV
one thousand	*kilo*volt	kV
one	volt	V
one thousandth	*milli*volt	mV
one millionth	*micro*volt	μV

Electrical symbols

The following table contains all the electrical symbols you are likely to meet:

symbol	meaning
————————	connecting wire-conductor
	conductors crossing, but insulated from and not touching on another
	junction of conductors
	double junction of conductors
	earth
	frame or chassis
	electric cell
	battery of cells-two or more cells connected together
	switch
	fixed resistor
	fixed resistor-alternative symbol
	variable resistor
	variable resistor-alternative symbol

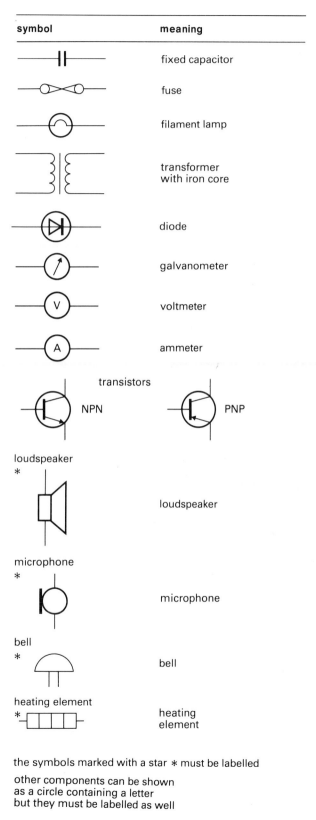

symbol	meaning
	fixed capacitor
	fuse
	filament lamp
	transformer with iron core
	diode
	galvanometer
	voltmeter
	ammeter

transistors

NPN   PNP

loudspeaker
*   loudspeaker

microphone
*   microphone

bell
*   bell

heating element
*   heating element

the symbols marked with a star * must be labelled

other components can be shown as a circle containing a letter but they must be labelled as well

Index

Index (cont.)